'Enlightening and inspiring. You will be thrown into reality and action towards an eco-friendly life by this book.'

— Cherry D.

'I loved the amusing anecdotes. The book is worth reading just for that.'

— David S.

'Living sustainably is not just good for the planet, it's good for your wellbeing and finances too; therefore, this book is a must-have for everybody, young and old.'

— V. Parmer

'A very enjoyable and well-written book. Charting a personal journey towards sustainability, it is bound to attract others who are keen to leave the world a better place for future generations.'

— T. Smith

'A fabulously written book. Some very sit up and take note moments beautifully put across. It has certainly inspired me to do better at my efforts but has also given me a feeling of love and hope. Together, we can do it!'

— D. Binyon

BEST OF ONE WORLD

60 steps to a sustainable, meaningful and joyful life

Hansa Pankhania

**BEST OF ONE WORLD is a sequel to the author's first memoir,
THE BEST OF THREE WORLDS, published in 2019**

Best of One World
60 steps to a sustainable, meaningful and joyful life
© Hansa Pankhania 2023

Published 2023
SOHUM PUBLICATIONS
Surrey, England

ISBN: 978-1-914201-17-2

A CIP catalogue record for this title is available
from the British Library.

The information given in this book should not be treated as a substitute
for professional medical advice; always consult a medical practitioner.
Any use of information in this book is at the reader's discretion and
risk. Neither the author nor the publisher can be held responsible
for any loss, claim or damage arising out of the use, or misuse, of the
suggestions made, the failure to take medical advice or for any material
on third party websites.

A billion tiny actions have brought us to the edge of an environmental crisis and a billion tiny actions can pull us back from the brink with small changes making a big difference collectively...

It is a great feeling when positive sustainable changes are seen to be made, but this path may not be easy at all times, as you will find out from my journey.

Nevertheless, we will still inch forward, just by taking the first step and gradually building the inches to feet and to yards and eventually miles that flow effortlessly, connecting all living beings into the vast oneness of the world.

CONTENTS

You and I have been given the historic responsibility to set things right. We have the unfathomable, great opportunity to be alive at the most decisive time in human history. The time has come for us all to tell the story and perhaps even change the ending.

– Greta Thunberg
Environmental Activist and Author

PROLOGUE

I am not an environmentalist nor a climate change activist. I am just a regular mother and grandmother who is very concerned about the future of her children, grandchildren and forthcoming generations. I also love nature and believe we are all connected. Therefore, a threat to other living species feels like a personal assault on me and my family. That is what inspired me to write this book. By writing and sharing my journey to stop these assaults, I feel I am taking control over what I can within this tiny corner of this vast wide world over which I feel I have no control. I have no control over the destruction of the Amazon rainforest or the destruction of marine life through industrial fishing. However, what I do have control of, is what I do or do not do in my daily life.

This is why I embarked on the path to an eco-friendlier lifestyle. **Taking these steps led me on an extraordinary journey through a world of joy I did *not* even know existed.**

In my first memoir, *The Best of Three Worlds*[1], published in 2019, I shared my journey highlighting the importance of unconditional love, psychological safety, forgiveness and other humanitarian values, as well as the fusion of the rich cultures and histories of three countries: Kenya, India and England. At that time, I was living in Birmingham, in the UK. The last lines in *The Best of Three Worlds* state, *'I wish with my heart that my future generations will capitalise on their inheritance of these family values, thirst for education and flair for entrepreneurship to make this a better and peaceful world for all living beings.'* I cannot expect them to do this if I am not doing everything within my power to make this a peaceful world for all living beings.

Following the publication of that book, 2020 was ablaze with major tumultuous events – Brexit, climate change, and most of all, worldwide lockdowns due to the Coronavirus rise in infections in February 2020 and global death tolls from this Covid-19 pandemic. This pandemic gripped the world and brought it to a standstill. We were confined in our homes, working or not working. We could not see our loved ones or rush around the country and the world as we were used to. As 2020 threw its door open to us, the bushfires in Australia were already rampaging, consuming 18.6 million hectares of land and some six thousand buildings. It was devastating for humans as well as wildlife. Three billion terrestrial vertebrates died in the fires, and some species may have been driven to extinction. Wildfires in the Brazilian Amazon and more acres of the world's 'green lungs' were devastated. Agricultural and industrial development exacerbated the wildfires on the west coast of the USA. The climate change debate gained momentum and there was a call from many directions for us to save trees or to use less water, energy and plastic[2]. In addition to this, the Black Lives Matter movement was fired up by the death of George Floyd. How can the life of a person not matter, just because of the colour of his skin?

It was a year of tremendous loss and gain. This collective turbulence and trauma meant we had to contend with much fear, anxiety and isolation but at the same time gain appreciation for our health and the support of loved ones. Remote working and being confined to our houses during lockdown triggered an awareness and appreciation of the nature around us. Empty supermarket shelves made us more aware of the food we eat and its distribution chain.

Did this pandemic change us as people? Did business become more socially and environmentally responsible? How did politics change? Did the lockdowns trigger the compassion within us that would save the planet? All these issues screamed at me and kept me pondering well into the night during that time.

It was a time when many of us were seeking answers and meaning in order to come to terms with these events. Why is this happening now? What are the lessons that humanity needs to learn from this suffering? The way I tried to make sense of this was the belief that the universe is angry. It has sent many messages via natural disasters for us to treat the planet more kindly, reduce the carbon footprint, reduce plastic, and take better care of all the species, but we did not listen enough. Eventually, it had to bring us to a standstill and confine us to our homes, so we would have to pause for a few months and have time to reflect on how we were living our lives. It enforced the changes that we should have voluntarily embraced in our lives many years ago, e.g. less flying and travelling, coming together as a community, and spending more time with family and friends.

As a child, whenever there was news of a natural disaster, my Baa (Mother) would say, *'Dharti ooper bhaar vathi gayo chhe,'* meaning Mother Earth is feeling weighed down.

Now Mother Earth was crying out loud but we were not listening or doing enough.

We are still not listening.

I am still not listening enough either.

The Covid crisis gave people much needed time to reflect, appreciate the basics of life and review their priorities. For me, that certainly happened as the questions on the scientific laws of nature and universal truths that life is based on, were revived. It feels as if I have been sleepwalking past this beautiful, stunning, intricate tapestry of nature all my life. However, I have now woken up from this slumber and started to notice the exquisite mosaic of nature around me. Like me, people started to appreciate how we live and care for our planet and make the connection between industrialisation, climate change and the demise of nature and its treasure of species. The realisation of how nature and every living species on the planet is crucial for our survival and the urgency for change started to creep into our obtuse psyches. We wanted to hold onto this tranquil feeling and commitment to preserve nature. However, few have since changed their behaviour permanently. To me, it feels as if the changes happened fleetingly. Is it because we got caught up in the rat race again, living with the belief that money means happiness and material possessions assign personal worth? But it is humbling to note that the richest people on earth were not immune to the Coronavirus infection, nor were they spared the implications of the wildfires and floods.

In January 2020, just before the Covid-19 pandemic, I moved to a village in Surrey to be nearer to my children and grandchildren. My new house is next to a vast meadow, approximately seventy metres from the River Wey, which runs parallel to Manor Farm and Manor House. Unlike the urban location of my house in Birmingham, there is an abundance of nature as soon as I set foot out of my front door. I have got into a routine of taking my daily walk along the riverbank and around Manor Farm. On my way back, I find a comfortable spot by the river and take in the magnificent nature around me to connect with the elements of Mother Earth, the reverence of the river and the soul of the sky, which guide me in the direction of my destiny and purpose in life. I get this feeling of complete oneness with the world, this feeling that we are all one entity; the people, flora and fauna, oceans, mountains, deserts, every living being in every continent of the world connected.

Then one day...

I am standing on the bridge over the weir, watching the white splashes as water gushes from under the bridge and takes its course further along the River Wey. My ears welcome the chirping and tweeting of the birds as they play with each other and fly from one tree to the next. A walker calls out to his dog in the meadow in the distance.

Suddenly this sweet natural orchestra turns into a cacophony.

Then there is silence.

Silence for a few long moments.

A little while later a mellifluous voice whispers in my ear and says, *'This paradise which is feasting your eyes and streaming its melody through your ears may not all be there for your granddaughters and their generations. The ecosystem of the planet is falling apart and one day, all this will be taken away from all of you!'*

My body, mind and spirit are all in chaos and then locked in stalemate, in that one breath that tries to make its way from my nose to my lungs. I break out in a sweat, my muscles flop like lifeless rags in my limbs, my heart chases its beats in desperation and the panorama of paradise before my eyes fades into a blank cold sheet of grey.

Although I am in agony, I manage to hold onto my breath as well as the picturesque vision that was in front of my eyes. I sink all my attention into that present moment and many more moments after. I stop searching for answers and simply make room for them in my heart. Gradually, I resurface from this state to the feel of my hands perched on the stone ledge. The rush of the river water jolts my senses back into my spirit. The birds go quiet.

The voice whispers again. *'Are you going to do anything about it? What are the small everyday actions and behaviour changes you need to make which will take the weight off our magnificent Mother Earth?'*

How can I have missed this message? The message being conveyed to me and all of us that we need to save our extraordinary planet. I see the bigger picture and hear about it in the news every day. How can I as one single pixel in this vast picture make any difference at all?

Mahatma Gandhi said, *'Be the change you want to be.'*

I have to change first, then encourage other pixels around me to make changes too. The more pixels I can reach out to, the more we can all change the bigger picture.

As a child, I lived a car-free, plastic-free, waste-free, vegetarian/almost vegan lifestyle, surrounded by beautiful natural landscapes and warm kind-hearted people. So, what changed that I lost the capacity to live that way? I moved from living in a small close-knit community, in a small town called Thika in Kenya, to the huge city of Birmingham in England. I got embroiled in consumerism, convenience foods and commuting by car. I lost the intrinsic relationship with nature. I was in this state for fifty years until my move to a village in Surrey, back into a small close-knit community.

A thought bolts out of my head like an arrow from a bow. *Can I recreate the same sustainable life that I lived as a child here?*

After all, some of the conditions are the same. I can endeavour to complete the circle of being brought up in a small town on a plant-based diet in tune with the laws of nature by revisiting the values of my childhood and creating a more sustainable lifestyle.

How about if we all did the same in our local environments?

A collective shift in values is imperative and needs to ripple across the globe, causing a seismic epiphany. We need to be forced to question the narrowness of our attitudes and to think of a new way of living, a new order for our priorities where material wealth slips to the bottom of the list and planetary wealth springs to the top. Is there a stir in our conscience after being exposed to the television images of floods and wildfires around the world, the growing coverage of climate activism and the groundbreaking work of pioneers such as David Attenborough?

Are **you** listening and doing something about this?

Am **I** listening and doing something about this?

I keep pondering over this for many more dark, silhouetted nights because I am writing this book at a time in history when the people in my close circle have heard about the climate crisis but the **commitment for change is not there**. By that I mean, the majority of people in my close circle are eating meat and dairy products, buying all their plastic-laden produce from supermarkets, driving petrol and diesel cars, and taking water and utilities for granted. No one is a vegan and only a small minority are vegetarian. There are no local shops around me where I can buy loose, plastic-free vegetables, groceries, toiletries, and cosmetics. The government has announced its energy strategy which still includes investment in fossil fuels. If we are to listen to the science, then living sustainably should be mainstream by now, but we are nowhere near that.

We are carrying on abusing the planet when science is telling us, loud and clear, that we are in a climate crisis.
We have to adopt a new sustainable way of life.
We have to change our mindset and our behaviour.
It is not a matter of choice anymore!

People are becoming aware that they need to do something to offset the damage, but do not know where to start and how to keep going. The intention of this book is to kick-start your journey and help you along it. As I mentioned at the start of this book, the journey may not be easy at all times, as you will find out from this book, but we will still inch forward and gradually the inches will become feet, yards and eventually miles, connecting all living beings into the vast oneness of the world.

I must admit I am a novice when it comes to knowing about the magic of nature and its infinite creative energy. When I started to write this book, I could count the names of trees and birds that I knew on one hand. Therefore, this journey I am sharing with you may appear shocking to others who have travelled much further. Our levels of awareness and ability to make changes will be different, but unless we make a commitment to do something, the transformation needed to protect the diversity of life on our planet will not take place.

Instead of having 'climate anxiety' where you feel fearful and powerless, this book aims to hand the power and control over to you, helping you to map a plan of action, assisting in getting you out of the inertia of inaction and instead propelling you into constructive action.

What we do as individuals may feel like a drop in the ocean, but I have faith that it will make a huge difference. My Baa always said, *'Tipe tipe sarower bandhai,'* that is, *An ocean is created one drop of water at a time.* Yes, governments and corporations have to take big action and I am shouting at them through my books and work to stop the downward spiral to a doomed planet, but we as individuals also need to do our bit to sustain that flow, otherwise, we will end up in a doomed place again. This is not about just saving the planet, but sustaining a healthy, robust, thriving planet for future generations. We all have to come together and do what we can, otherwise, we will let them down. I cringe as I think of my beautiful granddaughters and their future. I have to embark on a new path.

I set myself a challenge.

Then I set myself another challenge of sharing my journey in a book to inspire others to join me.

In this book, I share and bare my thoughts and actions on critical climate change presenting issues and exploration of some solutions and ideas. I add information and facts from both an Eastern and Western perspective. The East does not have all the answers, nor has the West, therefore I will try and present a coming together of philosophies. So, here is my journey to save what I can of our exquisite planet and hope you will join and support me on this path. I hope by sharing my ignorance, fears and challenges, I shall make your path easier and faster.

I am mapping my 60 steps to sustainable living. If the whole world did this (eight billion x 60), we would be taking four hundred billion steps together! Imagine the world after that! Unlike *Extinction Rebellion* and *Insulate Britain* using disruptive measures which can be effective in their own way, I aim to channel my frustration and concerns in an alternative way, through this book.

This is that journey! Won't you join me?

CHAPTER 1

WRITING FOR THE PLANET

In my lifetime I have witnessed a terrible decline...
in yours, you could and should witness a wonderful recovery.

– Sir David Attenborough
Broadcaster, Natural Historian and People's Advocate
COP26

I arrive with nine other people, including Gilly, Gary and Leanne. We are all hoping to be inspired by the tranquil countryside location and each other. We choose our rooms and space in the communal living room and conservatory, for our lone writing during the day. In the evenings, we can come together to share our ideas and inspirations. It is a self-catering retreat. We are provided with fresh eggs and vegetables from the kitchen garden. We are expected to extend our creativity to turning these into healthy delicious meals.

This book is born here, in this cottage, nestled within the glory of the Cotswolds in the UK. I have been keeping a diary and gathering my research for this book for some time, but to find time amid a busy life and make a start with it has been difficult, so I have put this life on hold by taking myself away on a writing retreat.

On day one, I make a start. Before I boot up my laptop or pick up my pen, I engage with my breathing and pivot to my heart centre. I ask for guidance from my angels and spirit guides. I feel a wave of emotion and know that I am loved and supported as I try to overcome my challenges and navigate through my endeavours to make a difference to other people and the planet.

I take a long mindful breath and start typing. My fingers tap on the keyboard while attempting to make sense of the filing cabinets of information stacked away in my brain. It all comes out as a haphazard jumble, with no structure or sense, but I carry on, fully trusting that at some point it will flow itself into content that will have meaning to me and my readers.

The next morning, I am sitting in the conservatory, the space that I have assigned myself for writing, watching the birds in the birdhouse through the window. A robin and sparrow chirp away, darting in and out of the barbed wire enclosure. The Cotswolds landscape rolls up to the horizon. The sun is pouring its peach and orange shades through the grey clouds, not feeling shy to show its grandeur that warms up the still morning air.

A white butterfly flutters from one window to the next, trying to entice me away from my writing with its alabaster beauty.

I feel a complete oneness with the world; every sense, every bone, every part of my body and mind in tune with my spirit. Still and at peace with all of creation. I tap on the keys of my laptop and let the words flow down from my mind to my fingers onto the Word document on my screen. The word count creeps up and up with the days on the retreat.

The building is next to a working farm. On the third morning, I watch the sheep as they graze, and have stimulating conversations about world issues, including sustainability, with colleagues as we go for walks in the countryside. We pass by an apple tree and pick a few. Gilly tells me how to make a quick crumble for dessert by using the sugar-free granola in the cupboard. At dinner time, she peels and chops the apples, adds a little brown sugar and cooks it for a few minutes until soft, then scoops it into bowls, sprinkling with granola and chopped almonds.

'We have saved energy worth twenty minutes of cooking time that a crumble would have taken in the oven,' she remarks with triumph.

The next morning, I join Gary on the terrace for breakfast. We can hear the faint sound of a tractor in the next field, chugging through the grass and sense the dew protesting as the wheels of the tractor squash its delicate drops of moisture into the thirsty soil. I turn my attention away from the dew to Gary, who is spreading his toast with marmalade in copious amounts. Gary loves marmalade and jam. In fact, he tells me, he cannot imagine life without marmalade. He has almost scoffed the entire jar by midweek, spreading thick layers on his breakfast toast. He tells me he gets through four jars a month!

My enthusiasm on the topic of sustainability takes over so I say, 'If you made your own, you would save all those jars. If more of us did that, we would save more jars and even waste less fruit. It is quite easy. I can show you.'

This is my first 'lecture' of the morning. I am inclined to give them from time to time just because I cannot contain the passion for my work.

So, later that afternoon we go blackberry picking along the hedges of the fields and also pick a few cooking apples from the apple tree in the garden. We are on the rota to cook dinner this evening, so I show him how to make his own jam.

I pile the grated apple and washed blackberries to fill a bowl, heat a pan on the stove, add the fruit and three-quarters of the same bowl of brown sugar and let it simmer for thirty minutes. I show him how to check when the jam has set, by cooling a small spoonful on a saucer and running a finger to check the firmness. I boil water in a separate pan and drop his empty, washed jam jar and lid into it for three minutes to sterilise it. When the jam has cooled a little, I fill the jar and seal it with the lid.

He is thrilled with the result.

'I can use up leftover strawberries, pick blackberries in the summer and make enough jam to last a year. One jar of jam costs three pounds average, for a mid-range quality. If I have on average forty-five a year, it would cost me £135. The sugar and energy would cost me about £35. I reckon I will potentially save £100. I am saving money, and the planet is better off without forty-five jars that need delivering and recycling in a year. Over ten years, that's four hundred and fifty jars. And if twenty more people made their own jam, that's nine hundred jars that accumulate over the years. If one hundred people...'

He goes on. Yes, Gary, we get the point!

The following day I am writing in the conservatory and the caw of a crow disturbs my train of thought. I look up and see its black body and piercing eyes watching me. Did you know, crows can recognise a human face? I feel it looking at me and I wonder if it will remember my face and go and tell the others in its group that are making a din beyond the fence. Leanne walks in and amidst a riveting chat about crows, tells me a group of crows is called a 'murder'. I find this hard to believe but it is the truth.

The next day, Leanne and I go for a walk through the vast lush green fields. The sheep in the adjacent field ignore us and carry on bleating about in the grass. The conversation slips from sheep to cows to eating beef and the harm it does to the planet.

'I read some facts about cows and methane gas two years ago and stopped eating meat. I was already cutting down on red meat and only eating chicken and fish. I loved my beef steaks but once I read those statistics... goodness me... it repulses me even if a think of a steak,' she tells me.

I cannot comprehend this as I have been raised as a vegetarian.

'I have that information on my laptop somewhere so I can send it to you when we get back,' she offers.

She emails me the info about meat consumption and how that is destroying our planet.

'Every year 7.6 billion humans chomp through meat from an amazing 65 billion animals. Raising these animals, and getting them from the farm to a burger bun, puts pressure on the planet in three ways.

'Firstly, almost a third of all usable land is devoted to raising animals for meat and milk. It's not just fields of grass playing home to cows and sheep, almost a third of crops we grow are to feed animals. This farmland has replaced forests and destroyed natural habitats. Secondly, livestock emits the same amount of greenhouse gases as all fossil-fuel guzzling vehicles in the world. Cows are the worst culprits. As they digest grass, they constantly burp and fart methane, one of the most harmful greenhouse gases. Each of the 1.5 billion cattle on the planet can expel up to 500 litres of methane in a day! Lastly, looking after all those animals accounts for nearly ten per cent of all freshwater that humans use each year. Even more water is used to keep meat clean and cool on its journey from field to fork. Animal farming is also one of the main sources of water pollution, because animal poo and other chemicals wash into rivers and oceans.' [1]

There is more.

'Half a leg of lamb equals 39.2 kg of carbon dioxide, or 150 km in an average car.

'Four large beef burgers, 27 kg of carbon dioxide, or 105 km in an average car.

'Twenty pork sausages, 12.1 kg of carbon dioxide, or 50 km in an average car.

'By 2050 at least 9 billion people will be sharing the planet as opposed to 3.5 billion in 2000. It will not be possible to feast on meat at the current levels. If we avoid eating meat and cheese for just one day a week, it could help the planet more than taking your family car off the road for five weeks.'[2]

These statistics rattle and ram my internal organs. I have to grab the chair from my standing position and sit down for a few minutes to steady myself.

I heard that scientists are trying to come up with artificial meat substitutes but in the meantime, I shall carry on enjoying my yummy bean burgers and hold onto hope and my chair!

I am thinking of bean burgers as I pack, ready to return home. I am feeling sad as the retreat draws to a close. It has been an inspiring, enjoyable and productive week. This retreat has not just been one where we have gained inspiration for our writing but also where we have exchanged tips and information to take us all a few steps forward on our journey as authors. It has been invigorating to spend the week with like-minded people in a peaceful, countryside location away from the hustle and bustle of the industrialised and polluted world.

The energy level in my body is bountiful, my mind has the clarity of a mirror and my spirit shines and shimmers away like the sun in the crystal sky through the window of the room I shall soon be leaving behind. The oneness of the world wraps around me, making me feel snug and safe just like the feel of my peach dressing gown on my skin.

Later, I am on the train home, reliving the week's experience all over again in the confines of my mind. I feel a surge of zing and zest in my being. I am a few more steps forward on my journey to sustainable living and writing. Life has a newness about its meaning.

PART 1

NATURE

Rivers give us so much more than water. They have flowed through our lives as surely as they have flowed through places. But as we have learned to harness their power, have we also forgotten to revere them?

— Willem Dafoe, Actor and
Robert MacFarlane, Writer and Naturalist

I shall never forget the first early morning walk by the river two days after we were in our Surrey home. It awakened something deep within me, making these words flood out through my fingers on the keyboard on my return home.

The specs of pink are getting larger in the silvery clouds and the birds begin to sing their first melodies of the day. The cool breeze caresses my face of the last traces of sleep from the bygone night. My feet leave a trail in the damp grass as they lead me to the river to the right of my new abode. My ears eagerly invite the soft murmur of the flowing water as my eyes take in the lush green of the trees that defend and preserve the sanctity of the River Wey. The yellow disc of the sun comes in view, awakening even more life into the earth. The heavens open up around me and in my heart, flooding it with joy and tranquillity.

Nature is free. There are no costs or side effects of this wonderful energy that we are bestowed with all around us. It makes sense to me to make full use of it to refresh and rejuvenate my physical, mental and spiritual wellbeing after my gruelling house move and the uncertainty of the pandemic.

Research on Positive Psychology in 2020[1] shows that even a short amount of exercise in nature has significant returns. Interestingly, exercising in the presence of water – e.g. walking or running by a river, the sea or doing water-based activities – generates even greater benefits.

'I will come and see you and spend time with you as much as I can,' I say to the River Wey as I revel in my walk, along what I call the muddy path that runs close to the riverbank on my estate. The path is lined with thick vegetation, shrubs and trees that are home to a range of species. I make this my version of forest bathing, as I step mindfully along the path, taking in my green surroundings together with the sounds of the river with all my senses.

Working from home predominantly gives me more time to enjoy my wild surroundings as well as my garden. I can have my short breaks from work, pottering in the garden or simply sitting silently and being present with nature.

It is one such moment at 7.00 a.m. at the height of summer. The heatwave we are going through is making it impossible to venture in the garden at midday as it is already 26 degrees centigrade.

I am perched on one of the garden furniture chairs, sipping my first cup of tea with leisure. The parakeets are having a slanging match and darting from the silver birch to the oak tree branches on the green adjoining my garden, under a blue sky peeping through a fine net of clouds. I spot the moon saying goodbye to the night as it sneaks away into the clouds. The blackbirds try to stay out of the way of my view, darting from one branch to the next. I see a squirrel slide up and into a hole in another branch on the oak.

When I cast my eyes down to the ground, I am surrounded by vibrant pink and indigo echinacea. A courgette, laying peacefully in one of my plants in my vegetable patch, watches all this commotion while my eyes feast on the scene of the trees, flowers and vegetables, my ears basking in the chitchat amongst the birds.

I feel at complete peace with myself and this amazing world. I marvel at the beauty of the universe, and my resolve to put my arms around this alluring beauty and protect it for my future generations becomes stronger than ever. My months of frustration and quest for answers as to my purpose in life bears fruition.

I stay still in these moments of utter inner peace and oneness.

I recall how the Buddha describes this profound experience. *'Even though the spokes of a wheel are turning around like crazy, I am still like the axis in the middle, in control of which direction I want my wheels to go.'*

CHAPTER 2

THE LIVING MAGIC OF NATURE

Each day brings the excitement of a new dawn, a new adventure, discovery, new opportunities, new challenges, magic, surprises, fun, laughter and love.

I get into the habit of enjoying my first cup of tea in the garden to the welcoming notes of bird songs that heighten my senses and sooth my soul. My creative juices get in the flow and this is often when the best ideas surface, out of the morning brain fog.

I resolve to strengthen my relationship with the natural world. I decide to be more intrepid and cram as much nature as I can into my home and work life, endeavouring to learn something new each week about our wonderful natural world. I make a start by aiming to learn about the ecosystems and wildlife in my vicinity. I tune in to it with all my senses, breathe in the air, listen to the sounds of the birds, and feel the rough bark of the tree trunks as I walk past. I learn that exercising in nature helps to improve my health and fitness, both physical and mental.

I take my first step.

STEP 1 – BE CURIOUS ABOUT THE NATURE AROUND ME

I am embarrassed to say that my knowledge about nature is more limited than many of my readers holding this book, despite my progress so far. However, a chance conversation with someone triggers my curiosity and is increasing my knowledge about the wonders of nature considerably.

One summer day, I am again doing my daily walk through Manor Farm. In the distance, I see a woman gazing up at a tree, oblivious of the passers-by. As I approach her, I see a huge camera hanging sidewards from her shoulder. She glances my way just as I walk past, and I smile at her. She smiles back.

'Is there a bird up that tree?' I ask.

'I think I saw a kestrel but cannot be sure,' she replies.

'Are you a professional photographer?'

I introduce myself and she tells me her name is Katie. She tells me she has a passion for nature and likes to capture unusual images. I see her by the river and walking around the farm on my walks many times and a friendship develops over a period of time. I start to learn about kestrels and skylarks.

Prior to this encounter, my thoughts would rarely be in tune with my surroundings during my daily walks, flinging back and forth between faraway places and to ongoing work and family issues. I am now noticing a shift to being more present and interested in my immediate environment. When I tune into the colourful detail and natural splendour of my neighbourhood, I feel my insides weighing down with the gravity of regret. I have missed all this magnificence for almost fifty years! My pace slows down, keeping me rooted to the spot I am standing on now, lest I should miss any of the minute spectacle of natural life that may add to the weight of this regret in the future. I want to savour this beauty and magic and let it nourish every cell deep at the core of my psyche.

The next day I enjoy the beauty of buttercups, bluebells and hawthorn. I meet a neighbour by the area of the river we call the 'pond' because of the wide space for ducks and geese to swim. Today, the ducks and geese dive playfully at their leisure. My neighbour walks past adding to the animated audience of the inhabitants and theatrics of the pond. She points out the elegant looking ducks, telling me that they are called mandarin ducks. I find out more about these spectacular birds.

The male mandarin duck has the most elaborate and ornate plumage with distinctive long orange feathers on the side of the face, orange 'sails' on the back, and pale orange flanks. The female is dull by comparison with a grey head and white stripe behind the eye, a brown back and mottled flanks.

'They were introduced to the UK from China and have become established following escapes from captivity. The main population of mandarin ducks is in south, central and eastern England, but small numbers occur in Wales, Northern England and Scotland. These birds like lakes with plenty of overhanging trees and bushes.[1]

The next day I see a family of swans, a couple and five cygnets swimming in the river when I go for my daily walk. I do not have my phone with me but am lucky to have a young neighbour passing by who takes a photo of the swans and offers to send it to me. I bless him for his kindness. After my walk, I sit by the river and meditate to soak in the early morning sun, the tranquil rumbling of the river water and soft swaying of the trees behind me.

Nature wraps and enthrals me on my walks as well as from my office window. Magpies, swallows, wrens, crows and parakeets all perform their melodrama and try to outdo each other with their range of notes as I carry on work with my clients and writing for my books. As they float and dive up and down the oak and silver birch, then pass my window and above my roof, I feel flutters of joy and a deepening sense of connection with creativity at large.

All the plants and insects that surround me now brush past my soul, creating ripples well beyond my physical being. After all, we are one and the same energy.

As Paulo Coelho says in his wonderful book, *The Alchemist*, *'All creation is written by the same one hand.'* [2]

And as humans, we need to connect more with outer elements of sun, earth, water, wind and space to nurture and nourish the elements within our inner worlds.

He also says that, *'Anything on the face of the earth can reveal the history of all things, a page in a book, a person's hand, a turn of a card or the flight of the birds. Whatever the thing observed, one could find a connection with their experience of the moment, a way of penetrating into the soul of the world.'* [3]

STEP 2 – IMMERSE IN NATURE, EVEN WHEN INSIDE THE HOUSE

It is possible to engage in nature, even without stepping outside the house. Here are some ways that I enjoy nature from within my four walls.

My living room emulates a garden. Green curtains and cushions with patterns of leaves, a grass-coloured rug and paintings of flowers, waterfalls and rivers on the walls. For a birthday, my artist neighbour gifted me with paintings of the waterfalls where I grew up in Kenya, one of trees and another of the River Wey, which take pride on my walls as a mini gallery of her work, showcasing her affection for us and ours for her.

The brown sofas face the double patio, so I have full view of the garden and veteran oak tree beyond the low fence and gate that opens up to the estate green. I have a few houseplants dotted around the house, which add to the natural ambience.

In addition to the décor, the use of natural images and sounds gives me access to the energy of nature without stepping outside. Every morning when I do my yoga, I have natural images on the television or tune in to natural sounds. I listen to and sing along to mantras as I am going about my household task. Mantras are natural sounds that hold and give out positive vibrations. I do this at most of my breaks from work too. An image full of the vibrancy of nature can do wonders for my work performance so I change my laptop screen to a natural image and ensure I am facing the window when working at my desk. This may sound like a trivial matter but facing the outside world does improve my clarity and engagement when working.

STEP 3 – INTEGRATE NATURE WITH BUSINESS

Businesses might be surprised to learn that spending time in nature can boost productivity. Several studies have shown that being in nature or even looking at pictures of nature can improve cognitive function and increase employee productivity, while also lowering cortisol and blood pressure, making us healthier and more effective in our daily tasks.

One study even found that just adding plants to the office environment could increase productivity by fifteen per cent [4]. That is huge! And if simply adding some greenery to the office can produce those kinds of productivity boosts, think about how much more you could achieve if you went further and tried to incorporate some time in nature into your daily routine?

STEP 4 – ENJOY WALKING MEETINGS

I have an office at home and consulting rooms in the village, a five-minute walk from my house. In the past, I would hire meeting rooms at a corporate venue but now I invite colleagues for walking meetings. Instead of sitting at a desk in a drab windowless meeting room, we not only feel energised and rejuvenated, but also get a boost of Vitamin D and a little endorphin kick from moving around and soaking in our surroundings. For those of you who like to exercise but cannot make it to the gym, hitting the trail during your lunch hour would reap multiple rewards.

The other day, during a walking meeting with a colleague, we saw a family of swans glide along with us upstream in tune with our footsteps. They created ripples in the reflections of the sun and trees in the water. We watched these images gather back to their shapes as the swans moved on, before resuming our discussion on the Wellbeing Training Project for a pharmaceutical company in London. We passed by two fishermen sat on the bank, enjoying the swan spectacle, and I said my usual, 'Any luck today?' One of them replied, 'Not yet, love,' so we carried on along the path on the riverbank, through the gate leading through to Manor Farm.

'This is so much better than having a meeting in an uninspiring, minimalist, grey office in Weybridge like we used to,' my colleague remarked, smiling as he set his eyes on the bridle path ahead of us, along the river leading us towards the green bridge.

STEP 5 – ENJOY ALFRESCO LUNCHES

It can be the same for lunch. Rather than eating my lunch at the dining table, I take my lunch to the garden and eat outside. When I am dining out or having off-site work meetings, I endeavour to choose a restaurant with a leafy patio or the one which is proactive in reducing its carbon footprint. This can be an option for you too, if you seek out parks and other communal green areas around your work and find eco-friendly eating outlets.

CHAPTER 3

VEGGIE DELIGHT

Sometimes the most helpful remedies for the human condition are not packaged in a pill box, they're actually packaged in nature.

– Dr Roger Seheult
American Pulmonologist

Whenever anything comes up about sustainability in relation to food, one idea is always repeated: *grow your own*.

STEP 6 – GROW MY OWN VEGETABLES

Home-grown food when grown sustainably, has a much lower carbon footprint than that produced by conventional agriculture, because it is organic as well as right in the backyard. There is no transport of vegetables to market or you driving to the retail outlets. Vegetable gardens store more carbon when permanently planted with fruit trees, berries, and perennial vegetables interspersed with annuals to fill gaps so my garden plan would need to consider all this. I would love to have an orchard but space only allows a mini apple tree, which holds itself proudly in the right-hand side corner of my garden.

As for vegetables, the only one that Baa grew in our small five-foot square patch in my childhood home was courgettes and her favourite, marigold flowers.

I therefore start by sowing my first courgette seeds. Hubby digs a border on the right side of the garden and buys me some sustainable pots to build my vegetable patch.

I sow tomatoes, radishes, onions, peas, beans and herbs – parsley and sage.

As I am doing this, the song which features parsley and sage in its lyrics by Simon & Garfunkel, *Scarborough Fair*, begins to hum in my ears.

I am singing this aloud, getting the lyrics in the wrong order and annoying my neighbours at the same time, I guess. But the main outcome for me is to build on my home produce so I also sow coriander and fenugreek, two of the most common herbs used in Indian cooking.

I learn that if I plant herbs like coriander, dill and fennel, these will attract ladybirds. I use these herbs in my cooking most weeks, especially coriander, so I am surprised to learn that they control pests without the need for harmful pesticides. Mint, basil, lavender, chives and rosemary are also natural pesticides, so I shall aim to grow these next year.

Lettuce is easy to grow. It is worth having a go at growing this if nothing else, because of the importance of leafy greens in our diet. I did not realise you can grow celery, pineapple, leeks and lettuce from stalks, and avocados from its stones. My celery is surviving, and the avocado is sprouting its first root after I suspended it with toothpicks and immersed the lower part of the stone in water. Taking cuttings from your existing plants instead of buying new each year will save money as well as plastic.

I have a stock of plastic pots for seedlings that were left by the previous owner of this house, but I now also use biodegradable plant pots and toilet roll tubes which can be planted directly into the ground. Once broken, plastic seed trays can be harder to recycle, so it is worth investing in sustainable ones made from wood or bamboo instead. Up to half a billion plastic plant pots end up in landfill in the UK each year[1] so invest in a terracotta version which looks great on the patio if you have one. As well as using toilet roll inners, you can make square shapes out of old newspapers. Seedlings need more water when you do this, but the bonus is that they will take less time to sprout up.

I have started to collect lolly and ice cream sticks for labelling plants instead of plastic ones. You can buy bamboo labels, but there is no point spending money on these when it is easy to upcycle old wooden ice lolly sticks. I am not keen on ordering plants and garden materials online but prefer to look for plastic-free options at local garden centres or buy bare root plants or sprout them from seeds. Again, I get this sense of the magic of the oneness with the world feeling when I see white thready growth of the first roots appearing.

By summer my vegetable patch is thriving with courgettes, tomatoes, radishes, onions, peas, beans, parsley, sage, coriander and fenugreek. I have a good harvest of courgettes, beans and peas. The tomatoes have not ripened and the radishes I dig out are small, marble-shaped and hard, so I only use the radish greens. The parsley and sage have flourished but the coriander and fenugreek have not taken off at all this summer. However, what I cannot emphasise enough is the pleasure of eating lunch from my garden produce. Freshly picked off a plant or lightly stir-fried with soy sauce, the feeling is out of this world. I urge you to try and grow your vegetables and have this gripping, satisfying taste bud experience for yourself. The flavour of the courgette you buy from the supermarket, which has been mass produced, travelled across countries, wrapped in plastic and sat on the shelf for a few days, can in no way match the essence of the one where your eyes and hands have witnessed the seed being sown in the ground, the first sprouting shoots and flowers and the miniscule shape of the courgette maturing until your hands reach out to harvest it for your meal.

All this sounds like a lot of work as I am reminded by my friends and family. Nevertheless, when I see the first flower in my courgette plant and a teeny weeny cauliflower raising its head in the midst of its green foliage, the oneness with the universe's magic is mind-blowing. No matter how many times you have eaten beans, tomatoes, courgettes and all the vegetables, nothing tastes as good as what you've grown yourself. Healthy broad bean shoots are also edible and make a tasty addition to warm summer salads if gently steamed and covered with melted butter, delicately flavoured with garlic and a good grind of black pepper.

You'll get two crops for the price of one!

Many such heavenly moments are bestowed on me on summer days. I am sitting in the garden relishing a colourful plate of stir-fried vegetables, when a light green butterfly whizzes its wings past my eyes and settles itself next to the echinacea.

I feel blessed and happy beyond what words I can find today.

This visceral oneness with the world can stay as it is, without words for the time being.

Until the blackflies arrive.

On my return home from a working trip abroad, I dart into my garden to say hello to my vegetable and herb plants and am shocked to see that I have got blackflies on my broad bean plants.

Until now I thought my crop of broad beans, sown last autumn and having survived a winter, would remain free of blackfly, only to discover dense clusters of them surrounding the top-most foliage. Gardening lore has it that broad bean plants planted over winter are less likely to be attacked by blackfly than beans sown in the spring but there are years when this doesn't hold true – and this seems to be one of them. I had left my plants unchecked whilst away and so the blackfly multiplied rapidly, in no time at all, so the plants are now showing signs of shrivelled flowers and weakened foliage.

Much to my consternation, I learn that blackfly is especially attracted to broad beans, forming thick clusters around the new leaves at the top of the plants, the stems and on the underside of younger leaves. I turn to my faithful friend Google for help and the following information appears.

STEP 7 – LEARN ABOUT BLACKFLIES

Information from the Riverside Garden Centre says:

Prevention is better than control, so check plants which are likely to become infested from early spring onwards and use a brush to remove small clusters of blackflies before they increase and spread.

A gardener's trick is to pinch off the tips of broad bean plants before blackfly appear in order to remove a favourite feeding point. Not only does this address the problem, but it also encourages better growth and development of strong pods.

But before resorting to one of these, try spraying the affected plants with a simple mix of water with a few drops of washing-up liquid added. Or mix a few drops of thyme, clove, rosemary, and peppermint essential oils in a spray bottle with water. Spray this mixture all over your infested broad bean plants to kill the blackfly and their young.

Another method involves leaving three or four crushed garlic cloves to soak in the water before spraying it on. The theory is that insects don't like garlic so garlic in the water will repel them. This really does the trick. As with all these sprays, they have to be reapplied after rain.[2]

As per this guidance, I snip off the shoots and spray the plant with the garlic-infused water. I offer Hubby a green bean shoots salad accompaniment with our dinner with bated breath. I expect a reaction similar to the one with the dandelion fritters episode I talk about later in the book, but he surprises me instead.

'This is delicious. You must make it again,' he remarks.

I do make it again, many times after that.

CHAPTER 4

GARDEN GLORY

Life is a chance to do things your way, not the cheapest way, not the most popular way, and not how others think you should. And a very, very, very precious chance, at that.

– Mike Dooley
Entrepreneur, Speaker
and New York Times Bestselling Author

STEP 8 – AIM FOR A LOW-CARBON GARDEN

Nature and gardens are synonymous. In spring, as nature begins to bloom with its buoyancy, our attention and focus is diverted more to our gardens. I start reading up on ways to grow a low-carbon garden. My reading says a low-carbon garden buzzes with life, sparkles with water, and is packed with plants. If designed well, it can act as a carbon sink, actively combating climate change. Although I have dabbled with growing vegetables, I am a novice gardener embarking on a journey of discovering the joys and challenges of a low-carbon garden.

On one of my walks through the village, I see a man tending to a huge garden alive with vibrant wildflowers, bird houses and a pond so I strike up a conversation. He informs me that, 'Every time I plant a seed, mulch my soil, or let my grass grow long, I could actively be increasing the carbon my garden absorbs. We return all that carbon dioxide right back into the atmosphere by firing up a petrol-powered mower, filling pots with peat-based compost, or scattering artificial fertilisers. So, low-carbon gardening needs a twin approach of lowering my carbon emissions to neutral by gardening sustainably, then maximising the carbon my garden sequesters and stores, helping it actively combat climate change.'

I am made aware too often that gardening sustainably is a long journey, and it's fine to start small and work up to the bigger changes later.[1] The important thing is that I and everyone do something, rather than nothing at all. My challenge is that our house in Surrey has only a quarter of the garden space we had in our house in Birmingham. My back garden is seven metres by ten metres and at the front, we have a two-metre by three-metre space. Therefore any element of a low-carbon design will have to be borrowed and adapted to fit into that restricted space.

I start by buying Hubby a birdhouse for his birthday, sprinkling wildflower seeds in my front garden, and cutting down on mowing the grass in the back garden, then intending to build in other interventions to lower my carbon footprint each year.

STEP 9 – COLLECT RAINWATER

I have a big plastic bucket which I have started using as a water butt in my garden to collect rainwater instead of using the hose. By rainwater harvesting, I endeavour to keep our dependence on mains water supplies to a minimum. I also add a large container in the kitchen sink to save grey water. Every litre or gallon of mains water we use adds to our carbon footprint.

As well as saving water, my research indicates that rainwater is also better for my plants due to its lower pH content. The minerals that are sometimes found in mains water, especially in hard water areas, can raise the pH of the plant root zone, which can affect nutrient availability. Rainwater also contains traces of nitrates, essential for plant growth.

It makes me wonder if rainwater is good for my house plants too. Most plants tolerate unsoftened tap water but fluoridated municipal water may cause a build-up of fluoride in soil, which might eventually be detrimental to houseplants, especially those with long slender leaves, such as my spider plant. What I need to do is to leave chlorinated or fluoridated water out overnight at room temperature in an open vessel to let the fluoride and chlorine gas evaporate before watering plants the next day.

After water, I consider the options for soil. I ensure I buy peat-free compost. Hubby points out that it is more expensive than the normal compost but he is happy to pay more as digging up valuable peatlands is not good for our environment. I am yet to create my own plant feed out of stinging nettles or comfrey. Apparently, all I have to do is stick a bunch in a bucket of water for a week, stirring occasionally, then feed my plants for free. I guess gloves are a must, if I am to attempt this, when picking nettles.

STEP 10 – BUILD A FEDGE OR HEDGE

My eyes are drawn to the hawthorn and bramble hedges as I take my usual daily walk round Manor Farm. Hedges are one of the most eco-friendly types of garden or country boundary. Not only do hedges actively sequester and store carbon, they also provide nesting sites, food sources, and shelter for wildlife.

On my return to my desk, I switch on my laptop and while Googling hedges, stumble onto something called 'fedges'. There are several definitions of a fedge. The main one is that it is a cross between a fence and a hedge. It can be used as an informal boundary between areas of the garden. Another definition is that it is an alternative to a hedge or fence but combines the properties of both to create a living willow border. I like the one that describes a fedge as a fence built from waste wood and prunings that serves as somewhere to stack woody garden waste while it slowly rots down, providing shelter for wildlife.

When my grandchildren come over the following week, we gather surplus branches and twigs from the copses by the river and build a fedge because this provides windbreaks and shelter for local wildlife or a handy way to dispose of anything too woody to go on the compost heap.

As yet, I have not been composting waste so I save the vegetable clippings and the wonky marble-sized hard radishes to make a start with that.

STEP 11 – LEARN TO COMPOST

I turn to my constant friend Google and YouTube for videos on how to make my own compost because composting at home has multiple environmental benefits.

Compost bins make good use of green waste from the kitchen and garden; this avoids waste going into landfill where it releases methane, a greenhouse gas eighty times more powerful than carbon dioxide. Soil mulched with compost holds onto nutrients and rainwater better, meaning less need to water and feed your garden. An all-round winner!

I commence by putting aside four medium spare plant pots from the garage (which doubles up as a garden shed because I have no room in the garden), then saving fruit and vegetable peelings and cardboard loo rolls to devise my own system of composting and allow a few weeks of trial and error. I begin the process by saving all the fruit and veg peel and stubs in a container by the sink. I tear up any cardboard and add to this. At the end of the day, I walk out to the back garden, chuck this in a medium plant pot and layer this with soil from another plant pot next to it. It takes about two weeks to fill up the pot, which I move to the back right corner of the garden away from the patio set. I do the same with the second pot and that too goes at the back of the garden when full. By the time the third pot is full, the compost is ready for use in the first pot. By now, I can hear some of my readers saying, 'Why don't you just get a compost bin?!'

Okay, okay, I hear you. The reason is that I have a small garden and the huge compost bin would stick out like a sore thumb and would also be quite smelly. The whole idea is to use what I already have instead of buying new plastic stuff. This system is working for me. It is heartening to see my medium pots live happily and discreetly at the back of the garden, with all the wonderful micro organisms within it enjoying my crunchy flavourful, packed with nutrients, vegetable and fruit peelings. Every week I gather and compost on average four pounds of peelings and save these from going to landfill as well as about two loo rolls. That adds up to approximately two hundred pounds a year and one hundred loo rolls. Now, if we all did that, we could save tons and tons of waste going into landfill and the methane it would end up producing.

STEP 12 – USE ECO-FRIENDLY GARDEN MATERIALS

My writing and research on sustainable living spills itself out in conversations with friends and family. One weekend the family get together and my uncle, to my utter horror, expresses a wish to landscape his garden including the addition of concrete paving.

'How do I convince him not to?' To seek an answer to this thought, I scroll for information online on my phone.

'*Concrete has a high carbon cost, so minimising its use will help keep your carbon footprint down. The materials you use in your garden can have a detrimental effect on the environment. Hard landscaping in particular can notch up some seriously high carbon emissions. Materials for hard landscaping, especially when newly manufactured and transported from the other side of the world, can have a huge carbon footprint. Large expanses of paving reflect heat, especially at night, adding to poor air quality and the heat island effect in cities.*'[2]

I continue to put my point across as politely as I can, stooping over my phone and hunting through Google to show him the hard facts. He reads from my phone.

'*Cement, used to make concrete, contributes almost 1 kg of carbon dioxide for every 1 kg produced, although some is reabsorbed as concrete is exposed to the air. Bricks add 250 g per 1 kg, and every square metre of stone patio adds about 47.5 kg of carbon dioxide, depending on the stone.*'[3]

'Luckily there are plenty of eco-friendly alternatives available,' my uncle says and continues reading.

'*The building industry is currently exploring greener alternatives to conventional cement-based concrete, made from waste materials or biodegradable fibres such as Ferrock which is made from ninety per cent waste materials, which reacts with carbon dioxide to create iron carbonate, actively absorbing carbon from the atmosphere. There is also Timbercrete which is a reduced-cement concrete, substituting up to ten per cent with sawmill waste to make a material that's lighter than concrete, available as blocks, bricks, and pavers. Rammed earth is compacted subsoil or chalk packed between temporary panels which can be used for steps and walls...*'[4]

My uncle stops reading aloud for everyone, takes a long pause and asks, 'How about permeable materials like gravel? I have some of that already in my garden. A sustainably designed garden can still have patios and paths, if they're kept to a minimum and made from recycled ceramic gravel or reclaimed stone slabs.'

STEP 13 – UPCYCLE GARDEN FITTINGS

By now, Hubby is getting used to my sustainable living mission and offers his support.

'What Hansa is trying to say is to not buy anything new unless absolutely necessary. Don't buy materials unless you have to. Work with what you already have in the garden. How about second-hand garden tables and benches? Second-hand or reclaimed materials from salvage yards will have less of a carbon cost than new ones. They will also give your garden a unique feel and look.'

Meanwhile, I am thinking of something I have seen on the television this morning and divert the boring conversation about cement onto something entirely different. 'The facts also say to keep bees! They are essential for our ecosystem and will help pollinate your garden.'

I gather that this is something that is beyond the comprehension of most of us sitting around in our lounge from the 'Oh nos' and 'Bees sting!' comments that flood the room! I keep this on my radar for when we have mustered the courage to do so.

CHAPTER 5

IN THE WILD

Imagine a world where razing of forests for animal feed is replaced by a thriving ecosystem and the hum of a million species. Where oceans are abundant with sea life like turtles and whales instead of supertrawler boats, and coral reefs flourish under sparkling pristine waters.

— Greenpeace

STEP 14 – PLANT ANY TREE

I have always wanted to have an abundance of fruit trees in my garden but do not have the space. Even though I had a large garden in Birmingham, Hubby was very protective of his landscaped flower beds and rockery and just would not let me grow fruits and vegetables amongst those. I had to do with a damson tree that was already there when we moved in. In our current house, I stop listening to him and plant a mini apple tree.

Trees suck in carbon, storing it in their rich trunks and locking it in the ground. When you plant a tree in your garden, you help offset not just your own carbon footprint, but that of your children too. It's one of the most effective ways your garden can help fight climate change. It is estimated that on average, a single broadleaf tree stores 2.9 tonnes of carbon in its lifetime. [1]

STEP 15 – REDUCE MOWING THE LAWN

Many of my tiny actions start with re-educating Hubby! Sometimes it is easy to persuade him to collaborate and make the household changes I am trying to implement and other times, it is frustrating trying to make him understand why we need to do things differently. One of these is mowing the lawn. He had insisted on mowing our front lawn and that of our immediate neighbours every time the grass was more than three inches long until the time I sprinkled wildflower seeds and attempted to persuade him to wait until they blossomed.

I love wildflowers and so does Hubby but not in his garden. In spring, our estate, fields and woodlands beyond are bursting with the vibrancy of snowdrops, daisies, buttercups, bluebells, and dandelions. I want to savour some of this tapestry of colours, white, yellow, purple and blue, from outside my front door and my patio.

'Let the grass grow and you'll be surprised how many flowers appear,' I say to him but it falls on deaf ears.

'Our garden will look like the beautiful meadow by the river with naturally occurring wildflowers which will attract clouds of butterflies, bees, and other insects that thrive there. We do not necessarily have to have a mown front garden so it looks gardened.'

This also falls on deaf ears.

'Please put away the mower and our lawn will revert to something resembling natural grassland – one of the most efficient carbon sinks. Wild lawns have the benefits of a grass lawn but without the eco-drawbacks of weekly mowing. Also, cutting the grass too short encourages moss. This then requires the effort, energy and possibly moss treatment chemicals to get rid of the problem!'

This still falls on deaf ears.

I give up, feeling frustrated. After all, for him it is the fear of 'What will the neighbours say?'

There are many delightful plants we can grow in a lawn instead of grass, such as Acaena. Tapestry lawns are full of flowers and low-growing, mat-forming plants. Other interesting plants include white clover, chamomile, creeping thyme and yarrow.

I read about rain gardens as well. A rain garden lies below the level of its surroundings, designed to absorb rainwater that runs off from a surface such as a patio or roof. Rain gardens and ponds are designed to absorb excess rainwater. Even a small garden can be turned into a rain garden but Hubby is having none of this.

I make a wish. One day my lawn will have all this.

There is one thing I am able to persuade him to do differently though, which is to choose longer living plants instead of annuals and bedding plants. This means less replanting and less soil cultivation, helping us avoid disturbing fragile ecosystems and releasing carbon that is locked in underground.

STEP 16 – LEARN ABOUT LOCAL WILDLIFE

In September 2022, our beloved Queen Elizabeth II passes away. The media is understandably full of this news, with the concerns and ramifications of the government's energy bill support plan fading away into oblivion for the time being. It takes me a few days for the news of her demise and the Ascension of King Charles III to sink in.

I am learning that nature is my saviour at low points like this so I turn to walking along the river to soothe my soul. As I pass the line of trees on the bridle path, a grey squirrel pops its head from behind an oak tree and scarpers up the trunk into the branches. Squirrels are soft, fluffy, cute creatures and I am comforted by this sight. However, there is also a raging debate about their introduction to this country from America, their impact on the red squirrel population and the environment as a whole. I reflect on the episode on the Surrey Hills, on the BBC's *Countryfile*,[2] where one group is going as far as introducing grey squirrel contraceptives to limit their population and the opposite group is setting up squirrel sanctuaries to preserve their natural right to exist!

A squirrel wanders into my back garden and we get all excited for my grandchildren. 'Oh, Ana and Surya, look! A squirrel on the garden fence!' Little do my tiny granddaughters know of the controversy surrounding these cute, fluffy creatures.

STEP 17 – FORAGE

The eco-living dialogue continues with a friend of mine one afternoon. She is a foodie and babbles away incessantly about food.

'How can we as individuals contribute to these ambitious and optimistic initiatives?' I ask, hoping she will engage with this topic. I am pleasantly surprised when she responds to the conversation about foraging.

I tell her, 'Locally we have abundant sweet chestnut trees, and hazelnut trees, Manor Farm even has plenty of blackberries, rosehips and apple trees – all free food available if you can just be bothered to collect!'

'Oh really? What else is emerging from your research?' she says.

'Okay. The main themes are: eat less or no meat, especially beef and dairy as cows emit methane which is worse than carbon dioxide; reduce food waste; and eat locally sourced seasonal food; drive and fly less; use renewable energy; make homes and businesses more efficient; have fewer children; cut back or ditch plastic; grow more trees; and enhance biodiversity.'

She does not respond for a while so I assume she is mulling over my spiel and wanting to engage with the issue.

'Was there something about eating better? How about we pick up on what you were saying earlier about eating seasonal foods?'

We are back to the conversation on food!

I tell her that I make changes to my diet according to the Ayurvedic Principles that guide my life as the seasons change. I start to ease off the spicy food at this time of the year, as the spring weather starts to provide the warmth of the spices to my body instead. I gently steer the conversation from food to the wonders of nature around us, hoping she will stay on track with the issue under discussion.

I trail off into a description of spring and how the blossoms on the meadow have laid out a carpet of their petals as I go on my daily walk. The oak outside my office window is sprouting its buds with pride and the silver birch fills up with its silvery shiny glory. I invite her to experience the magic of nature in my new surroundings.

'Shall we go for a walk?'

As we walk through Manor Farm, the Belted Galloway cows are mooing away merrily, robins rustling and swallows swaggering amongst the hedges. The dandelions are striking, laying a sun-coloured carpet on the pastures on both sides of the path we are walking through.

The conversation keeps steering onto food so I rummage through my brain to make a connection between the sights on our walk, food and carbon footprint.

'Did I tell you, I made dandelion fritters last week?'

Her response has much higher levels of animated energy than when I was talking about cows and birds.

'No way! Can you actually eat dandelions?'

'Yes, we can do more foraging to save travel miles. I have known that dandelion flowers are edible so on my walk last week I picked a few handfuls when I was out with my granddaughter. I let her gather the dandelion flowers. I have never tasted them so I looked up a recipe for dandelion flower fritters.'

Dandelions are highly nutritious, available almost everywhere. Do be sure, however, to only collect flower heads where you are certain no chemical sprays are used. Also avoid picking flowers near roadways because dandelions soak up all those car fumes. Lay them on a newspaper sheet outside for about ten or fifteen minutes so all the little insects crawl off them, then run cool water over them and pat them well dry. Next, mix gram flour, yoghurt, spices and a pinch of baking soda to make a batter. Dip a flower in the batter, coating it completely, and place it flower-side down in the hot oil in the skillet. Repeat until the skillet is half full of dipped flower heads. Cook until the batter is crispy, then turn the flowers over with tongs (or forks) and cook the opposite side. It's important to fully cook the batter on the flowers so it's crispy, not mushy.

'They were delightful. Perfect for a delicious lunch,' I say.

She stops walking and stares wide-eyed at the carpet of dandelions, and I am sure she is dreaming of a yummy plate of crunchy golden fritters.

After a long pause, she comes out of dreamland. 'Did your Hubby enjoy these too?'

'Hubby was horrified when I told him what I had for lunch. He gasped and said, "You can't eat those. They will make you sick!" The only way I could convince him of the nutritious benefits of dandelion was to sneak them in some pancakes a few days later! I blended a mix of flowers, water, spices and flour for the pancake batter. The day after he had devoured these, I shared the ingredients and reminded him that he had immensely enjoyed the pancakes after which he had not suffered any side effects. I have to come up with new tricks to get his buy-in for our eco-friendly life!'

Our laughter bellows through the fields, beyond the green bridge over the river.

STEP 18 – WALKS IN THE WILD

A month later, my neighbour announces she is feeling lethargic that day.

'My body did not want to leave the bed this morning. The sheets were akin to a large marshmallow, enveloping me with softness, sweetness and comfort.'

'I love your analogy,' I respond. 'It is a beautiful morning. Why don't we have a walk through Wisley? That might cheer you up?'

And so, we set off.

As we cross the River Wey over the green bridge, the sun dances through the trees and into clear reflections over the water of the river. Time flies as we walk and talk and soon we reach Wisley Common, a remarkable reserve at the junction of the M25 and the A3, eight-hundred acres of heathland and woodland.

I feel privileged to have this within half-an-hour's walk from my doorstep as it is a nationally important site for dragonflies and damselflies, with twenty species recorded. It also attracts many rare birds, including the Hobby, which is one of the few creatures that can actually catch dragonflies.

'This will be useful to add to your book. Did you know Surrey has lost eighty-five per cent of its heaths in the last two hundred years, so the Council and Surrey Wildlife Trust work hard to protect the remainder?'[3] she says.

'No, I did not. That is going in my book.'

We cross over the A3 bridge into Ockham Common, which is spectacular in summer. We take in the sight of purple heather, the beauty of the birches and the fragrance of Scots pines as we make our way to the Semaphore Tower on Chatley Heath.

'Here is another fact for your book. The tower was once part of a chain which was used to pass messages between the Admiralty in Whitehall and the Royal Naval Dockyard in Portsmouth. It was built in 1822 and is now the only restored surviving tower in a line of signalling stations that covered the entire route.'[4]

'That is fascinating. Also going in my book. Thanks, my dear.'

We continue enjoying the luscious landscape and stimulating chat on our way home and are soon back on our doorsteps.

'My tiredness has vanished. Instead, my body and soul feel renewed and nourished,' she tells me.

This is the nurturing magic of nature that brings all the elements of the world into one to help us heal and flourish.

I continue to cherish the wonders of the natural world in autumn and winter with my friends and family during our walks.

We savour the autumn reds of the Virginia tendrils that hang by the green bridge over the River Wey. We ponder over the toadstool mushrooms on the meadow, sweet chestnuts on the ground and in the woods beyond.

On mild winter days, I walk through Manor Farm, the site of which now consists of areas of wet meadow, improved grassland and a series of hedges, which attract species such as skylarks, pied wagtails, barn owls, linnets and roe deer. The dotted fan-foot moth – thought to be locally extinct in Surrey – has also been found on the reserve. The network of hedgerows create important natural corridors across the site, as well as providing food and shelter for insects, invertebrates, birds and small mammals.

I see a small grey and white bird with a white underbelly darting on the hedgerow as though putting on a show just for me. I think this is a pied wagtail and I resolve to look that up with the Surrey bird watch group. As I saunter through Manor Farm, the Belted Galloway cows completely ignore me and carry on grazing with their heads down to the ground.

I continue to immerse myself in the magnificence of nature.

For those of you who are feeling stressed, I would suggest you immerse yourself in nature too. Give yourself the opportunity to appreciate the tranquillity that will unleash your full potential, broaden your horizons and hook you into the soul of universal love and creativity.

Together, let's make our hearts sing with the birds and revel in the allure of nature around us. Ahh... My daily dose of nature taking me to that peaceful oneness with the world space deep in my core.

I am winning! And the world is feeling at one with this win too!

PART 2

WASTE AND WANT

We don't need a handful of people doing zero waste perfectly. We need millions of people doing it imperfectly.

– Anne-Marie Bonneau,
Zero-waste Chef

What is environmental waste? The U.S. Environmental Protection Agency (EPA) defines environmental waste as *any unnecessary resource use or release of substances into the water, land or air that could harm human health or the environment.*[1]

There are many forms of waste such as 'liquid waste' which is dirty water, washing water, organic liquids, waste detergents and sometimes rainwater.

'Solid rubbish' includes a large variety of items that may be found in households or commercial locations that comprises of recyclable rubbish or other hazardous waste.

'Household waste' is defined as *comprising of rubbish (such as bottles, cans, clothing, compost, disposables, food packaging, food scraps, newspapers and magazines, and garden trimmings) that originates from private homes, and also referred to as domestic or residential waste.*[2]

'Domestic waste' is produced in the home through everyday activities. Like all waste, it must be appropriately dealt with, otherwise domestic waste can affect the environment and impact human health. Domestic waste is split into categories: organic, toxic, recyclable, and soiled. Examples of organic waste are flowers, vegetables and fruit, kitchen waste and leaves. Toxic waste can be batteries, paints, old medicines and other chemicals. Recyclable waste includes glass, cardboard, paper, plastics, and metals, whereas soiled waste is things like nappies, cloth soiled with bodily fluids and animal waste.

The time taken for waste to break down varies considerably. Did you know organic waste only takes two weeks to disintegrate whereas a cigarette butt takes up to five years? An aluminium can takes up to five hundred years, plastic bags up to one thousand years, tyres up to two thousand years and glass up to a million years!! And aluminium foil in theory never disintegrates![3]

These figures are mind-blowing. I am utterly aghast to learn this and sense it will disturb you too as you take these facts in. As you would expect, the prevalence of non-biodegradable pollution has become a major environmental issue. With better technologies, we are producing durable materials that can withstand extreme temperatures. These materials are very useful, but create problems when it comes to dispersal. Non-biodegradable waste can't be broken down by microorganisms or the natural elements. As a result, non-biodegradable pollution is an environmental concern, threatening to overwhelm landfills.

Let's just take one item from all of the above: food. The UK throws away 9.5 million tonnes of food every year.[4] That's the highest amount in Europe – and it's even more problematic considering that over eight million UK residents are experiencing food poverty. Why does the UK produce so much food waste? There are a variety of causes, but the main ones are when shops and restaurants order more food than they need, consumers buy too much food, and consumers' failure to pay attention to expiry dates.

Unnecessary purchases contribute to domestic waste as well as wasting money. A 2019 report[5] found that the British public wasted £641 million shopping online and failing to return unwanted products. Packaging of food and other products account for a large proportion of waste. In the UK, seventy per cent of all plastic waste is single-use plastic packaging. Improper management of domestic waste is common, especially in urban areas of developing countries. The resulting waste build-up can negatively affect human health. It attracts disease-carrying insects, contaminates drinking water, and can cause air pollution.

A large quantity of domestic waste ends up in landfill sites. Landfill sites remove waste from homes and streets because it has to go somewhere! It's a cheap method of waste disposal. Waste in landfill has the potential to be burned as a source of energy but this is not common practice as yet so landfill sites contaminate the environment. The breakdown of waste produces landfill gas, a source of carbon emissions. These unsightly and smelly rubbish dumps contaminate the environment and threaten human health. Landfill sites produce a liquid called leachate, which can become toxic. If left unchecked, leachate can contaminate nearby waterways, negatively impacting aquatic ecosystems. Landfill sites have a sweet, sickly smell caused by landfill gases. These contain greenhouse gases and pollutants including methane, carbon dioxide, ammonia, and hydrogen sulphide. Short-term exposure to landfill gas can cause a range of health issues such as asthma, sleep problems, weight loss, and chest pain. Methane and carbon dioxide can affect the availability of oxygen to the tissues, resulting in coordination issues, fatigue, nausea, and unconsciousness. Prolonged exposure to landfill gases has been linked with cancer, respiratory disorders, and developmental defects in children.[6]

Once you have taken all this information in and have recovered from the shock, think about your day in the context of the above categories. How many times have you used a bin so far? My guess is, probably a lot! There is an astronomical volume of waste that we humans produce, and this comes from a wide array of activities including everything you chuck in your own bin!

CHAPTER 6

FOOD, GLORIOUS FOOD

Granting rights to animals is crucial for the moral wellbeing of the human race and we owe it to future generations to aim to bring compassion and peace to other species.

– Nitin Mehta MBE
Founder of the Young Indian Vegetarians Society,
and Animal Rights Campaigner

Today, I see a post on Instagram with a quote from Sir Paul and Linda McCartney: *'If slaughterhouses had glass around them, everyone would be vegetarian.'*[1]

Reducing the consumption of meat is supported at the same time by the UK government report released in May 2021 on National Food Strategy[2] which makes three main recommendations: more tax on salty and sugary processed food, eat less meat, and medical prescriptions for fruit and vegetables including healthy food education. The aim is to break the junk food cycle and reduce obesity and diet-related health issues, by replacing meat with vegetables and lentils and reducing sugar and salt consumption to save lives.

Talking about reducing certain items for consumption gets me thinking so I draw inspiration from Linda McCartney and try to give up eggs and dairy.

STEP 19 – ATTEMPT AT BECOMING A VEGAN

Once a year, I do a detox where I abstain from wheat, sugar, dairy and fried food. After my four-week detox recently, I begin to slowly introduce these foods back in my diet. This time, as I am adjusting back from detox month, I attempt to not go back to dairy and eggs so as to convert to a full vegan diet.

I express my frustrations to a friend who also has found it difficult to give up eggs and cheese.

'I am finding this very hard. I was not brought up to eat eggs and cheese but the taste has stubbornly lodged itself in my mouth after being introduced to these as part of a Western diet. I am dreaming of omelettes, cauliflower cheese and paneer (an Indian variation of cheese) koftas, and waking up yearning for these tastes. I was at my son's house at a party this weekend and everybody was enjoying all these delicious recipes with relish. It was hard to abstain.'

I have said before that no one in my immediate family is vegan except one niece of mine who is experimenting with it now, although a handful are vegetarians. My siblings like me were all brought up as vegetarians (or 'almost vegans,' meaning we had dairy milk but no eggs and cheese) in Kenya, but we were introduced to meat as part of the Western diet in England. My children made the personal choice to eat meat when their childminder offered that option to them.

In my attempt to give up eggs and cheese, I start to take vegan options for myself when I stay over at my children's. They have busy lives with young children, so I do not want to burden them with additional shopping and meal planning for my vegan options.

Sadly, I am not able to keep this up for more than a handful of weeks, but not for any of the above reasons. I miss cheese, although my first memory of the taste of cheese at a primary school function was of a soapy, textureless chunk in my mouth that I spat out because of its disgusting taste! Cheese manipulated itself over my taste buds and palate until I was hooked on it once I was in England.

Keeping to my commitment to reduce my carbon footprint, I try goat and sheep cheese, which do not quite make my palate zing with joy, but at least does not make it whinge and wiggle with dissatisfaction. There is no such drama with milk, however, as my taste buds are quite happy with soya, almond and oat milk except in my first cup of morning tea. For me, this was a good excuse to get used to black tea, for its health benefits. Black tea is rich in antioxidants that may provide benefits including improved heart and gut health, and lowered 'bad' LDL cholesterol, blood pressure, and blood sugar levels. According to Macrobiotic Health Coach Shilpa Arora, '*Milk makes tea acidic. Tea has potent antioxidants catechins and epicatechins, but adding milk cuts down the amount of these antioxidants making this otherwise healthy drink a source of inflammation and acidity.*'[3]

Black tea tastes bitter at first. I persevere and persevere until my taste buds stop throwing tantrums and accept that that is all they will get in the morning and for my mid-afternoon break.

I relive this drama with my taste buds when I try to give up cheese, eggs and dairy milk a few more times over the next months until one day, I am seduced back to the flavours of eggs. As for cheese, I try many brands of vegan cheese which my taste buds keep rejecting over and over again until the dairy cheese lures itself in my mouth from a platter of cheese and crackers at Christmas.

Now I continue to be an 'almost' vegan, having given up cow's milk, but not having the willpower to resist the random treat of an omelette, dairy cheese or yoghurt...

Especially yoghurt. Baa made copious use of yogurt in her appetizing recipes, my favourite being *Kadhi* (warm yoghurt soup). I want to stay with these sweet and savoury memories and keep talking about yoghurt for a little longer.

STEP 20 – MAKE MY OWN YOGHURT

I start to make my own yoghurt just as Baa made it, saving about fifty plastic yoghurt pots a year going into landfill. The process is quite simple. Bring two cups of milk to the boil and let it cool down to room temperature. Then add a heaped teaspoon of yoghurt as culture, mix well, cover and leave it in a warm place for eight to ten hours until it sets. I use a tea cosy over the pan to keep it warm, or you can leave it overnight by a warm radiator. Once it is set, it can go in the fridge. Save a small quantity as culture for the next batch. As well as saving money and carbon footprint, homemade yoghurt is healthier and lasts longer because it has no chemicals or added sugar and salt.

STEP 21 – CUT DOWN ON FOOD WASTE

My niece who is recently trying to become a vegan is visiting today. I have been working on my 'food' chapter this week so the conversation takes itself onto that topic.

I tell her, 'We were brought up not to waste any food at all. Any surplus food was distributed between neighbours as most of us did not have refrigerators, despite the heat of tropical Africa.'

She finds this hard to comprehend, being born and brought up in England.

'I still combat food waste by sharing with my neighbours or turning leftovers into lunches. I freeze leftovers, fresh herbs and even bananas to make them last longer.'

'Mum does all that except the bananas,' she tells me. 'Can you really freeze bananas? I'll have to give that a try.'

'I read in a book that you can even freeze eggs. Just beat them lightly and leave them in the freezer in a spare container. Also, I buy staples like rice and pulses in bulk to save money and reduce packaging or get them from the refill shops when I am in the vicinity.'

We go on to chat about homemade gifts always going down a treat. 'I am now making jams, chutneys, banana bread or cookies instead of giving loved ones shop-bought presents. Whenever something is close to its use by date, I look up recipes on YouTube. Today I will make frittata to use up leftover vegetables, eggs and cheese as an occasional treat.'

'How do you make that, Aunty?'

'I line a baking tray with brown paper and bread, fry onions, add chopped courgette, beans, spinach and pepper which are harvested from my garden, add beaten egg and grated parmesan and mix well, then pour this in the baking dish over the bread and plonk this in the oven for twenty minutes. I always try and use all the shelves in the oven to maximise energy use so I will cut up leftover fruit in thin slices and bake these on the spare shelves in the oven to make crisps. This makes a healthy and tasty snack. You are welcome to stay for dinner.'

She is thrilled. 'Hey, that's our afternoon snack and tonight's dinner sorted out for another day and no food waste in the bin!'

We give each other a high five.

I share more tips with her. 'Plan your meals for the week and stick to your shopping list, so you don't buy too much and reduce food waste. Make your weekly shop last longer by fermenting or pickling food. It's great for gut health and a flavoursome way to preserve it. Cut down on single-use drink cartons by blending your own juices and smoothies from leftover fresh fruit and vegetables.'

I pause while I tidy up some of the clutter on the kitchen worktop.

'Did you know that over three million tons of fruit and vegetables are wasted in the UK, before they even leave the farm? It's estimated that one in five bags of food shopping goes to waste each week. Across the country that adds up to a substantial amount of food, but also money wasted.'[4]

She is flabbergasted to learn that an average family of four could save £70 a month simply by reducing the amount of food they waste.

We look up the Surrey Environment Partnership site, which has lots of advice and tips on small, easy changes we can make to our daily habits to reduce the amount of food waste.

She reads aloud.

'*The issue is that over the last fifty years or so, our food system has become dominated by a handful of corporate supermarket supply chains. While they have delivered cheap food to millions, this convenience has come at enormous cost; to the environment through the promotion of unsustainable farming practices and high food miles; to local economies and communities through decimation of local food and farming businesses; and to health through the promotion of cheap, highly processed, nutritionally poor foods.*'[5]

I am glad one good thing came out of the Covid-19 pandemic. It has exposed the fragility of supermarket supply chains in terms of their capacity to respond to systemic shocks, resulting in empty shelves and a spike in food insecurity.

We look up more data on supporting locally sourced food outlets.

She reads aloud from the screen again.

'The Vocal for Local campaign is advocating for decentralised local and short supply chain food systems as the solution to many of the above problems and the route to a fairer, more sustainable, and more resilient food system that is fit to meet the climate, environmental and public health challenges that are becoming increasingly urgent.'[6]

I am curious and ask, 'Is there any specific ways that they suggest doing this?'

She continues to tap on the keyboard of my laptop, pause and read, tap at the keyboard again, read a little more and then after five minutes or so, exclaims, 'Ahh... I have found something...'

'Specifically, the campaign argues for explicit recognition of the failure of the supermarket supply system to deliver efficient distribution of tangible (food) and intangible (health of people and the environment) goods to all of society by acknowledging the potential of local and short supply chains to deliver just, sustainable, and resilient food systems that meet the challenges of the 21st century. It also aims to set National targets for local food production and distribution based on 80% domestic food production and 20% imports.'[7]

'Wow, those targets are pretty good,' I say with intrigue. 'But, how would they meet those?'

She continues reading aloud.

'A local food infrastructure fund that facilitates access to safe and nutritious food for at-risk populations and stimulates a thriving local food business economy by investing in cooperatively managed infrastructure. Dynamic public procurement models which source healthy, locally produced, seasonal and sustainable food to our schools, hospitals and care homes. Planning Policy and Local Business Rates applied to control the proliferation of supermarkets, while also creating conducive environments for social enterprise and independent food businesses on the other.'[8]

STEP 22 – SUPPORT LOCAL AND INDEPENDENT RETAIL OUTLETS

My niece and I carry on the debate about the challenges of sourcing locally produced food. There is only one farmers' market in my area, which is once a month on a Saturday and difficult to go to as that is when I am at my son's for grandkid duties. We do a search for local food producers, processors and distributors in my area, aiming to buy from local producers where possible. We locate a farm shop on the route to my daughter's house so I decide to pop in there on my return journey the following week.

My niece and I continue our online search in the meantime. The Vocal for Local website has pointers on hosting a Vocal for Local Food Feast! It says, *'Get together with friends, neighbours or people in your community to celebrate local food systems and producers over a plate of hearty local goodness. Take photos and upload to social media using the hashtag #VocalforLocalFeast to show your support for the campaign.'*[9]

At the present time in my life, most of my weekends are being taken up with looking after my grandkids, but I include this information for the benefit of other readers. I hope to run these events once the opportunity becomes available in a few years' time.

We continue to search and we find the information on food imports very enlightening. My niece reads: *'A shift away from imports to domestic production of vegetables in the UK would reduce GHG emissions by 7%! There is a strong link between local food systems and organic production. This is because farmers are able to keep a greater share of sale prices which they can then use to cover the increased costs of organic production.'*[10]

I add, 'Local produce also tends to be fresher than supermarket fruit and vegetables, as the distance travelled between field and consumer is minimal.'

'Listen to this, Aunty,' my niece continues. *'Research has also found that for every £1 invested in local food, between £6 and £8 is returned to society in the form of economic and social benefits, including training and skills, and health and wellbeing.'*[11]

'That is astounding.'

It just shows that there's power in every food purchase you make. The ingredients you buy show companies what people want, so if demand goes up for locally produced organic products that are good for the planet, they'll stop importing the bad ones. It's time for all of us to switch to local produce and spend wisely.

It's also time to stop food waste. The UK Energy Saving Trust say we waste an estimated 1.3 billion tonnes of food every year – around one third of all food produced for human consumption [12]. This scale of food waste leads to habitat destruction, decreased biodiversity and overuse of land and water. To combat the climate emergency and help protect our environment, we all need to reduce the amount of food we waste.

CHAPTER 7

DROPS OF WATER

What you do makes a difference.
And you have to decide what difference you want to make.

− Jane Goodall DBE
English Primatologist and Anthropologist

It is a January crisp, gold-filled day with frost on the grass, gleaming in the reflections of the mild winter sun. I am feeling elated on my walk, more than usual today, because of the wonders of the animal kingdom being bestowed on me with abundance. I encounter wrens and dunnocks darting from trees and hedgerows, two magnificent horses being ridden by chattering young girls on the bridle path along the river, a pair of deer who have switched pastures from further beyond to pastures near Manor House, the Belted Galloways close to the hedgerow on the field giving me piercing eye contact, and three geese paddling away by the pond near the weeping willow.

I feel blessed and at one within me, within the world and far beyond over the whole universe.

And then it starts to pelt down with rain and to sleet. Rainwater everywhere!

STEP 23 – SAVE WATER

Many things are going well on my reducing carbon footprint and sustainable journey. Then it's more good news.

I get a surprise incoming telephone call from my water provider. It is refreshing and encouraging to get a personal call from a service provider in contrast to the reams of information imparted by email from other providers. She introduces herself as a GetWaterFit Coach. The very friendly caller goes on to offer me a video conversation via Zoom where she advises on the Saving Our Streams scheme[1] and how to save water usage at home. She tells me, 'Because you have accepted and joined this video call with me, you'll receive two hundred and fifty token coins which you can then donate to planting trees to help reduce global carbon levels. This call will only take approximately fifteen minutes and could also help save you additional money on your bills. Not many people know this, but the water we drink comes from the local environment, flowing through nearby rivers and streams. Completing challenges on GetWaterFit will encourage you to save more water and in turn save money. When you complete a challenge and record your results in your dashboard, you'll earn up to one hundred and fifty coins each time. Simply log in each day and track your progress. Most challenges last three days and you can start more than one at a time. For every one thousand coins donated to Save Our Streams, our water company will plant a tree.'

We continue to talk about changing showerheads, having shorter showers, using water-saving gel in the garden and other water-saving tips that would assist in saving our streams. I am surprised to know from her that I can look up my water usage, compare it with average use and keep a record of my savings on my account. I do the Save Our Streams Challenge, earn coins and add to growing trees.

STEP 24 – REDUCE SHOWER AND BATH TIME

Thinking about showers takes me back to my childhood days and I reminisce on how we would have our daily cleansing makeshift shower. Every morning Baa would keep a supply of hot water on a wood burner which I would transfer to a pail and take to the *navni* (Gujarati name for bathroom). I would sit on a *patlo*, a low stool, wet the body, soap it and rinse it off by scooping water from the pail with a jug. That method used far less water than the contemporary showers where the water keeps gushing out even when you are soaping and scrubbing. The navni also gave a sense of freedom as it is an open space in a room, rather than the confined tight square of a shower cubicle. I think about trying this in my current morning routine, but my bathroom in the UK is not designed for that. Maybe bathroom designers can take note of this in support of the eco-friendly way of life. The bathroom is where we use the most water and by only using the water we need, the biggest savings could be made here.

I was dumbfounded when I discovered how much money all of us can save by getting into good habits that save water. The figures I share were set out succinctly on a Waters Worth Saving website[2]. All statistics are provided by the Energy Saving Trust[3] and are for a four-person gas-heated household. Water bill savings also apply if the household is on a water meter. How long do *you* spend in the shower? The average shower is eight minutes long and uses almost one hundred litres of hot water! Instead, by having four-minute showers, a household could save £165 a year on the energy bill, and an additional £100 on the water bill if on a water meter. After viewing the data, I get serious about reducing my shower time by working on a shower routine of three to four minutes maximum.

Even though as a child we did not have instant hot water, and heated water on a wood burner instead, I was brought up to always start the day with a shower and never leave the house without one. I have kept that up to now but after being enlightened about the benefits of saving water I occasionally, when I am working from home, swap the shower for a wash and wipe down. I could have made my shower even more efficient by having the right showerhead but when I called in our local plumber for advice, he said, 'Sorry, Mrs P, you have a large and fixed showerhead that can't be easily replaced without lots of other structural work.'

When I enquired about electric showers, he said, 'An aerated electric shower could damage the shower unit, and electric showers are already water efficient.'

You might be able to install an efficient showerhead known as an 'aerated' or 'regulated' showerhead. It saves on water without reducing the strength of flow, by mixing air into the water. A household could save £75 a year on energy bills and another £45 if on a water meter.

Not everyone has baths but if you do then they can add up to a big amount of water and energy, with the average bath using eighty litres. If you're having multiple baths each week and you're able to swap one for a four-minute shower instead, you could save £35 on energy and £18 on water. I often have a relaxing soak in the bath after a particularly hectic day, perhaps once a month. Knowing this triggers a bout of guilt after which I aim to reduce this by replacing the soak for a long meditation practice that would not require any waste of vital resources but refresh and relax me just the same.

STEP 25 – USE WASHING MACHINE AND DISHWASHER ONLY WHEN FULL

The washing machine is the kitchen appliance that uses the most water, and on average our household uses it just under ten times a month. Once I started putting it on only with a full load, that cut out two runs a month, and along with running it at an eco-setting of thirty degrees, I potentially save £68 on energy and £10 on water a year.[4]

A full dishwasher is often more water and energy efficient than washing by hand. The Indian way of washing-up is notorious for water wastage. I do not use a washing up bowl, but soap and rinse each item individually. In light of what I am learning now, it would be better to use the dishwasher on a full load more often than using running water over a sink full of utensils. My only frustration is that Hubby does not enjoy stacking the dishwasher and I am not keen on emptying it afterwards especially on a morning of an early work start when I am looking for my favourite cup or bowl in a hurry. I remind myself (and have to reinforce this to Hubby too) that only running it when full could help us cut out one run a week and save £17 on energy a year.

Just like the showerhead, an aerator fitted on the kitchen tap will save on water without reducing the strength of the flow, by mixing air into the water. If on a water meter, it could save a household £30 a year on water bills. When it comes to the use of a kettle, it's easy to overfill it, but by only filling it with the water you need, a saving of £13 on energy can be made.

These are additional actions where I shall need loving persuasive powers to encourage Hubby to make changes towards our environment friendly endeavours.

STEP 26 – TURN TAP OFF WHEN BRUSHING TEETH

I am happy to say that I took up this habit quite a while ago, but it's reminding Hubby and the grandkids to turn off taps when cleaning teeth and other household tasks. They are all getting more used to my 'Please turn off the tap while you do that,' and getting better at saving water.

I tell them that if they leave the tap off while brushing their teeth then this is one of the easiest changes to save water, because teeth really do not care if the tap is on or off while they are being cleaned. In fact, teeth would incur damage and begin to grit and grind after becoming aware of the fact that a running tap uses six litres a minute and it could save £40 a year if everyone in the house turned it off while brushing their teeth!

STEP 27 – HAND-WASH CLOTHES

Some bathrooms in the UK are designed in a way where the shower is over a bath tub. Since living in the UK, I have discovered an easy way to hand-wash the clothes not suitable for machine washing. I place the garment in the bath and as I shower, the soaking and soaping is done so I can rinse it after my shower and hang it on the shower rail for the water to drip before I put it out to dry.

Handwashing clothes always takes me back to life in Thika, Kenya. Baa used to hand-wash two pails of clothes every day in our outdoor *chokdi*, which was a cemented little square by a wall with a water tap. This scene where Baa is sitting on her patlo, bashing a garment with a *dhoko* (a bat like wooden stick) to release the dirt and grime from my brother's muddy trousers, is vivid in my mind. Then the scene switches to my neighbour doing the same in the house across the courtyard. I chuckle aloud as I am writing this, because my neighbour would be ranting aloud as she struck the dhoko on to the clothes repeatedly. It is only many years later when I was at university learning about anger management that I made sense of her behaviour. There you go. You have saved water and energy when you hand-wash in the bath when showering and you can also bash the garment about and let go of some anger!

Saving water is in everyone's interest. When it comes to clean water, it is so frustrating that virtually all water companies discharge raw sewage into rivers and the sea, tens of thousands of times a year, rather than invest in the infrastructure to treat it properly. Instead of contributing to the climate change issues in a positive way, they would sooner pay themselves huge salaries as well as making money for their shareholders!

With lots to worry about, it's easy to forget about water. But you can make savings by thinking about how often and how long you use hot water for. Only use the water you need and use appliances more carefully. This way you can cut out the water and energy you're wasting and so save money on your bills, as well as address climate change issues.

CHAPTER 8

MORE POWER WITH LESS ENERGY

Until you dream, there isn't a mould.
Until you speak, there isn't a promise.
And until you move, there isn't a path.

— Mike Dooley
Entrepreneur, Speaker
and New York Times Bestselling Author

Just as I reach a point of less turmoil on my sustainable living path, in February 2022 the whole world is thrown into chaos by the unprovoked attack on Ukraine by Russia. The months fly past with the news being dominated by the war and all its consequences. The fear of how this will impact on the availability of gas and energy grips the headlines.

The media shouts that global energy demand could rise by as much as fifty-eight per cent over the next three decades. To satisfy this demand and avoid a climate catastrophe, corporations and governments begin to switch the focus onto the possibility of renewable electricity. There's just one hitch: unlike coal, oil or natural gas, electricity is difficult to store, and renewables are primarily consumed close to where they are produced. Therefore, solving this problem will require stitching together the world's electric grids with the help of underwater cables.

'Grids are the forgotten giant of global energy investment,' says Alessandro Blasi, an advisor at the International Energy Agency[1]. 'Building new sources of power is important but not enough, as the power needs to reach consuming centres.'

'More than ever, it is necessary to have an electricity system even more interconnected at the international level,' says Stefano Donnarumma, CEO of international grid operator, Terna[2]. 'With increasing shares of renewables, it will be possible to contain and gradually reduce energy costs, to the benefit of the environment, citizens and businesses.'

There is also much discussion about energy efficiency in our homes.

STEP 28 – WRITE TO MY MP

I feel stunned to find out that homes in the UK currently account for around twenty per cent of our emissions, yet action to reduce these emissions has been slow, with 21 million homes currently below EPC rating C, meaning they are not efficient.[3] As the host of COP26 global climate talks in 2021, the UK must lead by example in cutting emissions, but our outdated housing stock is holding us back. We're currently joint last in Europe for sales of energy-efficient heat pumps and losing out on the green jobs they provide.

UK energy bills are high for a few reasons. Firstly, we have some of the worst insulated houses in Europe, meaning heat leaks from our homes, making us use more energy to keep warm. By upgrading and insulating our homes, households could save hundreds on energy bills. Secondly, we're still addicted to gas, which is nine times more expensive than renewables. By swapping boilers for heat pumps and rapidly expanding renewable energy, we could stop our dependence on volatile fossil fuels. These measures won't just save people money, they would tackle the climate crisis too. Thirdly, energy companies are making bumper profits and paying out dividends to executives and shareholders instead of pricing the public fairly.

At first, I feel there is nothing I can do as an individual to get this message through, but after speaking to like-minded friends, I write to my Member of Parliament to express my frustrations and concerns.

The government has a robust role in cutting out emissions without any doubt, but equally we all need to do what we can. Hence, I review my energy use at home and begin with a few more small steps.

STEP 29 – STOP IRONING CLOTHES

I recall a conversation with a friend who said she could not remember the last time she had used an iron, so I set myself the challenge of not using the iron for a month as an experiment, which leads to this tale of the crazy creases.

I was someone who ironed everything before wear. Skirts, tops, dresses, sarees, saree tops, nightdress and cardigans. Nothing got worn without first being warmed, caressed and smoothed over by the iron. Then one day I do it differently.

I have an appointment with the dentist. I pick a t-shirt and matching skirt from my wardrobe that have ample creases and deliberate whether to iron away the creases or not. I put them on, and the crinkles and creases stare at me, gaping lines in the folds of the cotton fabric, like a roughly beaten landscape.

I would be walking through the village to get to the dentist and wonder what judgement my fellow villagers will make as they see me breeze past in crinkly clothes. For that matter, whether the dentist in her chair next to me will be distracted from my teeth to the sharp folds on my top and flowing skirt. I take the risk and decide not to use the iron because my fellow villagers will probably be occupied in their own inner worlds to notice, and I trust my dentist with her professionalism to focus on the wellbeing of my teeth rather than my clothes!

Now I am sat in the dentist chair. While the young pleasant dentist works on sorting out my root canal, I take my thoughts away from the room and all its weird equipment to my memories of a holiday in Australia twenty years ago. I often do this in a dentist chair, that is, think about a holiday, a good book or a film to take my mind away from the drilling and poking that ensues. After an hour, I leave feeling relieved. Relieved because the treatment is over but relieved because nobody mentioned my crinkly, creased clothes. All this busy mental chatter in my head, I think on the walk back home, because of a tinge of vanity on my part and a tangled fear of being judged by others because of a few, actually quite a few, crazy creases on my clothes.

Electric clothes irons are normally rated between 500 watts to 2,000 watts. A 2,000-watt electric iron running for fifteen minutes every day will consume around 0.5 kWh of electricity in a day, and 15 kWh of electricity in a month. This would roughly translate to a monthly electricity cost of £5.40 in the UK. If I keep to my resolve and refuse to iron clothes unless absolutely, and I mean absolutely necessary, I shall save £60 a year on average. If we all started to do that, just imagine the collective saving we can potentially make? Also imagine how much life we are seeping away from our beautiful earth because of our vanity and fear of judgement from others? If we all let go of our vanity and fear of being judged by others in the same way, we can save colossal amounts of energy that could go in other directions such as food production.

On my way home from the dentist, I take a detour through Manor Farm. There is a renewed spring in the steps I take, as well as spring sprouting in all the trees, shrubs and wildflowers around me that adorn the meadow and pastures.

The whole world has sprung and come together into an electrifying oneness.

STEP 30 – STOP USING HAIR STRAIGHTENERS

The decision to not use a hair straightener came about with the 'cutting my hair at home' incident during the Covid-19 lockdown.

All hair salons were closed. My thick mop of hair was getting bulkier and longer. It started to irritate me until I had to take the risk of Hubby nervously clutching a pair of scissors to chop it off. He tried to emulate the environment of a salon, draped me with a towel on my shoulders, hair dampened, scissors at a mock professional level. Hubby snipped away and snipped a little more. This went on for ten minutes.

'I think I am done. Have a look in the mirror,' he said looking very pleased with his efforts.

'Oh no!' I was aghast. 'The right side is shorter than the left.'

'Is it? Okay, I'll level it up.'

Snipping and more snipping. I looked in the mirror after another five minutes.

'Oh, I don't believe this!'

'See. It looks lovely! That length really suits you,' he beamed.

I looked in the mirror again and burst out laughing. 'Ahhh… Now the left side is shorter than the right!'

'It's your thick and wavy hair. I just can't get it to level up,' he said as his laughter merged with mine.

After another action replay, he managed to balance the right and left hair lengths, but by this time, my hair was half the length it used to be. When it was long, I had to straighten it and clip it back every day to keep it under control. Now that it is this short, there is no need for that. I get compliments. 'Short hair really suits you.' 'Is this your new London look?' 'It makes you look younger.' Not only that, but it is also easier to wash and dry and I have no more use of hair straighteners, which means I am saving more energy for the planet.

I wonder what the villagers are thinking as I strut up and down our High Road to and from shops and bus stops, my wavy, wild hair flying and flopping about with each step. I remind myself that living sustainably is mostly about letting go of the fear of being judged and one's vanity. Therefore the hair straighteners stay tucked away in the back of my dressing table, grieving the loss of good times with my long, thick hair.

My calculation tells me I would save on average £50 a year for giving up on my faithful hair straighteners, averaging this out to £1 a week for fifty weeks a year. Now, if more of us did the same, would we not leave a better legacy for our future generations?

STEP 31 – SWITCH ENERGY PROVIDER TO RENEWABLE ONE

With the challenges faced of the high cost of raw fuel by small independent renewable energy companies after the Covid-19 pandemic and Russian invasion of Ukraine, many folded, including my provider. To my disappointment, a global fossil fuel company takes over my energy provision.

The global company is cheaper than the renewable one. I am in a quandary for a few weeks as the difference is considerable, and I could potentially make a good amount of savings. Eventually, all the money I save is for my children as their inheritance. However, my moral compass kicks in. I make the choice of transferring over to another renewable energy provider even though this costs far more. I would rather leave the legacy of a green planet as my inheritance instead of the money I would save from an unethical global company.

CHAPTER 9

CLEVER WITH CLOTHES

The existence of this world is simply a guarantee
that there exists a world that is perfect.
The world is created so that its visible objects
and humans could understand its marvel and wisdom.

– Paulo Coelho
Author of The Alchemist

STEP 32 – REPURPOSE OLD GARMENTS AND MEND CLOTHES

As with food, we can all reduce clothes waste. I pick up a needle and thread after years and take to darning socks and mending vests and dresses. Prior to my interest in sustainable living, these would have ended up in my bin and onto landfill.

On average, I am rescuing one garment per month, which makes for twelve a year. Just visualise the difference it would make if we all did this? I am proud to say, I removed a leather elasticated waistband which had ripped on a skirt, turned in the fabric and inserted another piece of elastic in it all by handsewing. Every time I wear it, my face lights up with pride for the planet. I also repurposed frayed towels into flannels and an old skirt into aprons.

I am beaming as I share my achievements with my sister.

'Where do you find all this time? You write books, run a wellbeing consultancy, look after your grandchildren, make homemade food, grow your vegetables and now you are handsewing and mending clothes!'

'These are tasks I find relaxing as I am listening to the radio or even watching television. If you have a passion and an intention for something, time will present itself to you. Your passion will make adjustments into your day and sneak in when you are least expecting it.'

I continue to repurpose and mend clothes instead of buying new ones, which I have not done since starting to write this book.

I am sorting out light clothing for an anticipated ongoing heatwave and come across an old linen blue dress, now faded in colour after several summers of wear. I have only one very light cotton nightie, preferring soft warm ones for a cosy night's sleep but these would be too warm for the heatwave nights.

'Ahh... I can use this as a nightie. It's linen, soft and comfortable. Nobody has made any rules about not wearing a dress for sleeping as long as you get a comfortable night's sleep,' I hear myself say. That night, in my 'new' nightie, I sleep like a log, as The Beatles' song says, although it did not feel as if I had been working like a dog! I had been working all day, but when your work is stimulating and aligned to your true purpose, it does not feel like work but rather a labour of love and a source of immense joy.

The next day, I cut off the frayed bottom of a pair of trousers to make shorts and undo the sleeves of a light top for comfort in the heatwave. I save the sleeves in the sewing box should I need them for later or could use them for craft play with my grandkids. I do not have to buy any extra new clothes, or any from the second-hand shop. I am ready for the heatwave without spending a single penny. I make a personal saving, and carbon footprint for the planet too.

STEP 33 – BUY SECOND-HAND CLOTHES

I also stop buying clothes made of man-made and synthetic materials. Instead, I opt for clothes that are made with natural fibres from the second-hand shop. I had never considered the option of buying clothes from a charity shop until the sustainability issues changed my way of thinking. I am endeavouring not to buy any new clothes unless in exceptional circumstances.

STEP 34 – SEW OWN CLOTHES WHEN POSSIBLE

After a few weeks, I meet my sister for coffee in the village.
'What's new?' I ask her
'I have joined sewing classes.'
'That is quite an extension of your creative flair.'
'How is your sewing going?' she asks me.
'As you know at one time, I could sew but these skills have got rusty over the years. I gave away my sewing machine to a friend in the belief that I would not have time to sew anymore with my other commitments. I can sew and did use to sew dresses for my daughter when she was a baby. I also had a go with saree blouses, because these are expensive to be sewn by a tailor. A matching blouse can turn out to be more expensive than the saree itself and can be about £30 but that is a skill I did not master.'
'Now that you are not buying new clothes, maybe you can have a go at making your own,' she suggests.

'I guess I can start by making simple things like straight skirts, but there are items I would have to buy or have them made like saree blouses.'

There is a pause. She is pondering something, and I wonder if she is thinking about the same thing.

Then we both hear ourselves say excitedly in tandem, 'Who says you cannot make your own saree blouses! Let's have a saree blouse-making afternoon together!' So, our next rainy day activity together is sorted and we have saved the planet a little bit more by not posting fabrics to India for sewing.

We continue our chat on ethical wardrobes.

STEP 35 – CUT DOWN ON WASHING AND SPINNING CLOTHES

You may not know that your wardrobe is a hidden source of plastic pollution. Many clothes are made from polyester and other synthetic materials which shed plastic microfibres, which then causes plastic pollution in our oceans. 'The average lifecycle of a garment is just 3.3 years, but the average polyester product is likely to last two hundred years in landfill. From the summit of Mount Everest to the depths of the oceans, microplastics have been found across the planet. Shockingly, they've even been detected in human lungs and blood. A recent study found that there have been twelve types of microplastic detected, which are commonly found in packaging, bottles, clothing, rope and twine, and many manufacturing processes.'[1]

'My goodness, that is horrifying. What can we *do* about this?' says my sister.

'We can stop our wardrobe contributing to plastic pollution and impacting our health by changing how we wash, care for and shop for clothes,' I tell her. 'Wash your clothes less. Do you really need to wash that pair of jeans or jumper after just a couple of wears? We're in the habit of chucking yesterday's clothes in the laundry basket but airing your clothes may well do the trick – they'll last longer and will save you money in the long run too. Spin less. Check if you can reduce the spin setting on your washing machine or, if you have the time, consider handwashing instead to minimise microfibre shedding.'

'That's great, Hansa. Keep nagging me about these things so I stay on top of it,' she replies.

'Sure, I will. That's a promise. There is nothing more I enjoy than to have a nag at you occasionally!'

Our giggling reverberates out loud, much to the amusement of other coffee drinkers in the café.

STEP 36 – ORGANISE A CLOTHES-SWAPPING EVENT

Clothes swapping with friends is a great way to keep your wardrobe fresh and save money, while avoiding fast-fashion and plastic packaging.

I grew up wearing my sister's hand-me-downs and swapping was common in our close-knit community where school uniforms, sarees, dresses, shirts and trousers were passed over from one family to the other. Luckily, since my daughter became a teenager, we have remained more or less the same size and have exchanged clothes, as well as shoes and accessories. It is common practice between female members of our family to swap sarees and other party clothes after having worn them a few times, especially the expensive silk and chiffon ones. I feel sad that the wider society has lost sight of this in our modern heavy-consumer culture.

Although we have been carrying out this practice informally between friends and family, I have never organised a formal clothes swapping event called 'swishing' – a fun and free way of updating our wardrobe without putting strain on the planet.

I add this to my future 'to-do' list.

CHAPTER 10

PAPER IS PRECIOUS

We do not inherit the Earth from our ancestors;
we borrow it from our children.

– Native American Proverb

STEP 37 – STOP BUYING CHRISTMAS CARDS

At the end of my walk on a certain day, I step foot inside my house and catch the pile of birthday cards on the floor that the postman has delivered while I have been out. This takes my thoughts back to Christmas when I was attempting to cut down on cards and gift wrap paper. It was the same dilemma when the Christmas cards plonked through the letterbox last year.

'I don't know what to do.' I seek solutions from Hubby. 'I could reciprocate with more cards as expected or run the risk of being seen as anti-social or even rude for not giving cards to my near and dear friends and family.' It is no good asking a man who has just bought a large pack of cards to give to his colleagues at work.

'What's the big deal? It is just a Christmas card after all. Everyone at work is giving these out.'

How do I get through to people and not sound as if I am preaching?

My heart begins to sink so I take a few deep breaths to keep it afloat.

It *is* a big deal for me. I churn over it for a few days and eventually inform my friends, family and neighbours that I am not giving out Christmas cards this year and will wish them a happy festive season personally or by telephone.

Having resolved the card issue, the next predicament looms in the glittery rolls of wrapping paper that Hubby has purchased with the pack of cards.

STEP 38 – CUT DOWN ON WRAPPING PAPER

When I was growing up, gifts were wrapped in brown paper, with a string. There was no tradition of birthday or cards for special occasions. Instead, personal visits were made to pass on best wishes and celebrate special occasions when possible.

'I got some brown paper but will my presents look dull and cheap wrapped in it?'

I turn to Hubby again for support.

'We do this every year. Why can't we do what we do every year? You are being really weird all of a sudden,' he says.

Habits are not easy to break. I discover that this journey to sustainable living will have many bumps, twists and turns. Just getting Hubby to see my point of view is like climbing a steep hill and I have yet to conquer the mountains on my path ahead. I can't even convince one person so how am I going to convince the whole world?

There have been many heart-sinking moments on this path and this is one of them.

In the weeks leading up to Christmas, the cards keep pouring in, and gifts wrapped in non-recyclable paper with lots of plastic trimmings also get delivered. The email I get from Greenpeace highlights a lack of awareness of a sustainable Christmas. It says our plastic is exported to Turkey and other countries for recycling but when that does not happen, it contributes to pollution there.

My chest and belly cramp up with feelings of sadness and frustration. The changes towards an eco-friendly life should be gaining momentum, but it seems this is happening only at a snail's pace. Change needs to happen at home first. I am gently getting through to Hubby. He means well and always does his best to support me although his responses may be abrupt at times when he is tired or preoccupied with his work issues. I hold on stubbornly to my commitment of not sending paper cards and try e-cards instead.

We go shopping for brown paper and recyclable wrapping paper. It is not readily available and after scanning the major supermarkets, we manage to get some in a homewares store. I have still got plenty of the shiny non-recyclable wrapping paper left from last year. I wrap the presents in brown paper or recycled boxes and add a strip of the shiny paper to add interest and colour. I make small flower shapes from the trimmed edges of the shiny leftover paper, for decoration. I feel much happier with this compromise and my family and friends are impressed!

This year at Christmas, Hubby begins to get it. He has asked me to make jam and chutneys from our garden produce as meaningful gifts in contrast to shop-bought commercial plastic and paper-laden gifts. When my granddaughter comes for a sleepover, we make lots of homemade cards with scrap cardboard and paper where she draws images and fills them with vibrant colours, adding, 'Merry Christmas' in her innocent and cute handwriting.

'Everyone at work has given me a card and pressie,' he says, 'and I shall have great pleasure in reciprocating with our homemade gifts and cards. From April 2023 first class postage for a card or letter will be £1.15. Trying to live sustainably is not only good for the planet but for our pockets too!' he exclaims.

We are winning and as we all pull together on this mission, the world is winning into oneness too.

CHAPTER 11

THE ROAD LESS TRAVELLED

There must be a better way to make the things we want,
a way that doesn't spoil the sky, or the rain or the land.

– Sir Paul McCartney
Musician and Writer

STEP 39 – REDUCE USE OF PERSONAL CAR

During the Covid-19 lockdown, I had reduced the use of my car and taken to more walking. Not everyone can take this drastic step for practical reasons but one day after much heart searching and wrenching, I take a leap of faith. I sell my car as the next step towards reducing my carbon footprint.

I get flutters in my belly and my heart feels weighted down with the loss of my car, which has been a reliable constant companion, a loyal friend which enabled the journeys with my loved ones and accompanied me to all four corners of the country for my work. I learned to drive forty years ago and since then had got used to hopping into my car for small as well as long journeys. My work involves travel to businesses nationally, so it was not unusual for me to set off in my car at 6.00 a.m. for a training programme or a set of coaching sessions, a journey over four to five hours away to the opposite end of the country. I would sometimes stay in a hotel depending on how many days of work I had with a certain organisation and then drive back the same distance after a busy working day.

So now the car is sold, I have to experiment and adjust to using public transport, which at first fills me with trepidation. To help me overcome this and seek my next steps, I look up my local borough website. The stats show that transport accounts for thirty per cent of the borough's carbon footprint. The borough is taking initiatives to reduce this, firstly by encouraging walking, cycling and public transport as the first choice for short trips.

STEP 40 – GET THE BETTERPOINTS APP[1]

Surrey County Council has teamed up with the BetterPoints Active Travel app to offer rewards every time a car journey is switched for a more sustainable mode of travel.

BetterPoints are points that I can use to reward myself by exchanging points for high street vouchers, or I can donate to charities or community groups. Whether I walk, or catch a bus, every journey counts, so I download the BetterPoints app and start earning points today. For every one thousand points, I can redeem £1 at a range of retailers.

STEP 41 – LEARN TO CYCLE

Learning to cycle has never entered my head, simply because of my faithful car and me behind the even more faithful steering wheel, forging my bygone journeys.

Now I have no car.

A thought sneaks in my head. *I need to learn to cycle now.*

Another one shouts back, *For goodness' sake, you are sixty-seven years old. It is too late.*

The first one responds, *People learn new things in their nineties, so sixty-seven is not too late.* Then yells, *It's good for your cardiovascular health and wellbeing overall.*

The second one protests, *I do enough of that in other ways.*

My thoughts have an ensuing battle and the one which says, *Sixty-seven is too late*, wins for the time being. No cycling for me. I am too old; it is far too dangerous and the infrastructure is not in place yet to reassure me of my safety.

I cannot let you win, the other thought fights back. *Literally, get on your bike and get cycling. Look up local lessons now!*

I obey and the information on the Bikeability website[2] reassures and calms the wrangle in my mind. It says:

If you're worried about getting started, we want you to know it's not that unusual. Eighty per cent of adults rarely cycle, if ever, and this increases for women and ethnic minorities.

Learning to cycle doesn't take as long as you might think! Our instructors can get the most nervous riders pedalling in under an hour.

You won't have to learn with children. You will complete your adult course either one-to-one or in a small group of adults.

You don't even need your own cycle. We offer free minutes for all Bikeability learners, so you can learn before you buy.

I take the bold step of dropping them an email seeking information on local lessons. My thoughts stop bickering as they await the response with tenuous anticipation.

(P.S. Due to work and granny-duty commitments, I did not get round to starting the bike lessons sooner but have my first one in the week I am handing in my final manuscript for this book to my publisher. I shall leave you to speculate on my progress. It could be that I abandoned the lessons for various reasons or that a petite Indian-looking lady with a helmet riding a bike is a constant sight in a little village in Surrey!)

In the meantime, it's buses, trains, trams, taxis and walking for me.

STEP 42 – USE MORE PUBLIC TRANSPORT

One flashback jolts out of my mind when I think about buses. My memory of taking a bus when we were a one-car family during the 1980s is where I am running for the bus in my stilettos. Most workday mornings, I would be late for the bus after getting two children ready for school and dropping them to the childminder. The shopkeepers by the bus stop would watch with bated breath, expecting me to fall over while running for the bus. But they never got the pleasure of seeing this. You see, I had mastered the art of running for the bus in my stilettos, briefcase in one hand and handbag dangling in the other as I jumped on the bus just as the driver was about to take off.

The time on the bus was a welcome relief from the hectic morning routine. I would do my breathing and relaxation exercises to help me ground for the day. The walk to work after the journey in the fresh air kickstarted my brain cells and physical body for the challenging day's work ahead.

We did become a two-car family soon after falling into the trap of the collective middle-class belief, 'Oh, but we need two cars because of our work, which is at different times and in a different direction.'

I chose not to see that there were many buses and trains in this different direction. Being stuck in traffic and concentrating on the driving, struggling to find a parking space and paying much more for parking than bus fares did not put my brain and body into gear for the day's work ahead. In fact the journey would be stressful, where I would often need half the morning for recovery. I had to rely on a walk at lunchtime (if time allowed) for my brain cells to be fully receptive to the demands at work. With this lack of exercise, I would probably be tempted to join a gym, but if I continued to get the bus, I would get all the exercise as well as fresh air, which would save the drive to and from the gym and cost of my gym subscription, as well as provide exposure to nature on my journey, time to relax and time to watch the world go by.

This is living proof that getting the bus is good for our wellbeing, for work efficiency, for our purse and the wellbeing of the planet in many ways. For the record, I have never joined a gym, despite trying it a few times, for the same reasons. I prefer my daily routine of half an hour of yoga in the morning, a thirty to fifty minute walk in nature where possible, and twice a day fifteen-minute meditation. This amounts to between ninety to one hundred and twenty minutes a day which is the same time it would take if I went to the gym and all without the monetary cost and significant damage to the planet in travel miles.

I would love to tell you what it was like getting on the bus for the first time after years of driving everywhere.

I have a meeting with a colleague in Woking and am standing at my village bus stop with my bus pass clutched anxiously in my hand. I rack my brains to recall my last bus journey. It was about twenty years ago. It was when you had to have the right change which you dropped in the machine by the driver, in exchange for a ticket. Prior to that you had bus conductors who would keep an eye on each new passenger and do the rounds for collecting the fares.

I share my apprehension with a lady who is also waiting for the same bus. She tells me, 'It is all digital now. You press your pass or bank card on the icon on the machine to register your fare.'

I thank her for her guidance. I see the bus appearing from down the road and it starts to halt as it gets to my stop. I get on and tap my card the wrong way on the machine initially but get it right after instructions from the driver.

I watch the world go by as the bus struts along.

Since then, there have been many such enjoyable journeys until the void of not having my faithful car slowly fills up. To get my daily dose of nature, I take a detour and walk along the river and through the Manor Farm on my way to and from the bus stop.

As they say, one step forward and two steps back.

'What is that 'something' that will encourage people towards using public transport?' I ask Hubby. 'It's good for the environment.' I suggest to Hubby that he needs to get used to using public transport as he is far too much in love with his car.

'No way, I am not going on the train or bus,' he protests.

I search my brain for when he last used a bus or train and no memory of this comes alive. He loves his car. It is his pride and joy.

'We have to come together as a community, have more car sharing schemes, give neighbours lifts to the station, and explore sharing school runs for our grandkids. We have to let go of our English reserve and strike up these conversations. If you know or see someone dropping kids off at the same time at school, volunteer to share cars for pickups and drop offs.'

'I told you, I am not sharing my car or hopping on a train. I love listening to my music and having my own space on my car journey'.

'But you can listen to your music on the train too.'

That has not happened as yet. I'll let him enjoy his car of course but also vow to get him on a train or bus very soon!

STEP 43 – WORK FROM HOME

I guess it is easy for me to say this since my switch to working from home recently. One great thing that came about during the pandemic lockdown is that our commutes were deemed unnecessary some of the time. It gave rise to the 'hybrid working model.' Homeworking for all the week or partial week started to become the norm, so we're not all commuting at the same time anymore.

Many of my clients made the adjustment to remote sessions. Zoom meetings zoomed up in popularity. Flexible work is now a fact of life: ninety-three per cent of workers globally want the freedom to decide where and when they do their job. Most people who took up homeworking because of the Coronavirus pandemic plan to both work from home and in the workplace (hybrid work) in the future, according to data from the Opinions and Lifestyle Survey (OPN)[3].

Workers were asked about their future plans in February 2022, after government guidance to work from home when possible was lifted in England and Scotland. More than eight in ten workers who had to work from home during the Coronavirus pandemic said they planned to hybrid work.

'To improve our quality of life, we need to become less dependent on mobility and more committed to local proximity. Making work, even in an office, just a walk or short bike ride away may be in store for more of us. Barcelona's Superblocks and the fifteen-minute city in Paris offer hints at what's to come,' says Pieter Cranenbroek, Senior News Editor at LinkedIn.[4]

PART 3

PLASTIC

Life without plentiful plastic
(Oh, those sweet, plastic-free days!)

*The greatest threat to our planet is
the belief that someone else will save it.*

— Robert Swan
Author, Public Speaker, Polar Explorer,
and Transformational Leader in Sustainability

On the surface it feels as if this is just an ordinary evening on a June day in 2019 after a working day. However, something extraordinary happens in my internal and external world.

I am flicking through the television channels absent-mindedly. My eyes are drawn to a programme titled *War on Plastic*[1] presented by Hugh Fearnley-Whittingstall and Anita Rani. I press the select button and after that, I am riveted for the next hour. My insides go topsy-turvy as I watch the programme.

Plastic, plastic, plastic everywhere. I am petrified by how much plastic we produce, use, and discard and the subsequent

101

damage it causes to all life on the planet. Even our best efforts at recycling are harming the world. We are exporting the plastic waste to Far East countries like Malaysia. In the piles of plastic waste exported from the UK, there are bags from Sainsbury's, Tesco, and other major supermarkets.

I go on the internet to find out more and as I read, my stomach starts to knot and as I carry on reading, the knots twist and turn and get tighter. The fear for my grandkids and their future mangles into a tight grip in my stomach.

These apprehensions rob my sleep for the night. After a while, instead of lying awake worrying about my grandchildren and their future generations, I strengthen my intention to do everything I possibly can to make this a less plastic world.

Corporates and governments must have the right strategies in place to address this serious issue too. Big and small brands and supermarkets need to step up and provide reusable alternatives that everyone can use and afford. They have the power to make the big changes that are needed. But that doesn't mean individuals aren't powerful too: the changes we make to reduce our own plastic footprint can make an astronomical difference.

CHAPTER 12

PUT A STOP TO LITTER

There is no such thing as "away".
When we throw anything away it must go somewhere.

– Annie Leonard
Proponent of Sustainability and
Executive Director, Greenpeace USA

STEP 44 – DO A PLASTIC COUNT

I am getting regular updates and information from Greenpeace. One of them is about the big plastic count and so I start counting.

To reduce litter, I have to muster up the courage to find out how much I produce in the first place.

This is my breakdown of the plastic items I gather over a week in July 2019.

One big bottle, one small bottle, three fruit and veg trays, three snack wrappers, two drink packages, four soft food and drink packaging, one deodorant tube, one bleach bottle, one toothpaste tube, two split plastic bags, and one eye pencil, which in total, is twenty pieces. Over a year this would average out at one thousand pieces. I would say we are an average user who have made some small changes but have a long way to go. But can you perceive the total number for the world population?

A further breakdown of my plastic use was seventy per cent food and drink, fifteen per cent toiletries and fifteen per cent other.

With the Greenpeace Plastic Count initiative in July 2022, almost 250,000 people took part and together they counted a shocking 6.4 million pieces of plastic.[1] On average, each household threw away sixty-six pieces of plastic packaging per week, which is 3,432 pieces over a year. Applied to the UK as a whole, UK households throw away an unbelievable 1.85 billion pieces of plastic a week or 96.6 billion pieces of plastic a year.

I would urge you to do your own count. How does it compare with the above?

Greenpeace tells us that out of this, only eleven per cent is recycled in the UK, sixteen per cent is exported, twenty-five per cent goes to landfill and forty-eight per cent is incinerated.[2]

I am perturbed! What can I do? There is simply too much plastic being produced and our recycling systems cannot cope, so we are burying it underground or sending it to Third World countries to clear our mess. With plastic production set to double over the next twenty years, we all need to make representation to our politicians for drastic measures to be put in place to reduce this.

When I share experiences with my circle of people, they responded by saying it was 'overwhelming,' they were 'very sad and frustrated,' and they would 'have to now take responsibility for my own overuse.'

Is this a case of too little, too late?

We are in 2023 and only a few days ago, I saw a media report of plastic bags in the stomach of a whale. We are reminded regularly that our beautiful planet is paying the price of our overuse of plastic and will continue to do so for decades to come.

Let's all aim to bring it down week by week.

STEP 45 – BUY A MULTIUSE WATER BOTTLE

I stop buying single-use bottles of water and use a reusable one. I fill it up with fresh water at a start of a journey. I have a small and large multiuse one for when out and about on foot or on public transport.

I reflect more on plastic bottles and my childhood plastic-free days.

There were no plastic single-use water bottles. We had water fountains in school and carried water in a stainless-steel sealed container on the rare occasions when we travelled to Nairobi. Now I see single-use plastic bottles everywhere – restaurants, homes, cars, pavements, meadows and woodlands. I even see them on busy roundabouts or traffic light signals when a driver or passer-by has had the discourtesy and lack of manners to throw one out of a car or when walking about. When I see this, the blood begins to simmer in my feet and continues to heat up, the temperature increasing until it gets to boiling point in my head.

This earth that is those pavements, meadows, woodlands and the grass verges by the roundabouts, is home to millions of species (8.7 million in the world in fact) of living creatures and organisms. How would you feel if someone flung rubbish in your home? Worse still, if you mistook this for lifesaving food and consumed it? I say to all those who chuck plastic waste or for that matter any sort of waste on the roadside, fields or anywhere except litter bins, to think again. It does not take much effort to keep it with you until you get home or near a litter bin. Just as we want to keep our home healthy and a safe, rubbish-free environment, so does every other living being on earth. Please, please share a thought for them and keep your rubbish to yourself.

STEP 46 – PICK UP LITTER

In the three and a half years I have lived by the River Wey, I have often seen litter strewn along the riverbed and especially in an area in the middle of our estate we call the 'beach', as this has a wider space for sitting or having a waddle in the river. My thoughts on seeing the litter there? *Stupid, irresponsible people. Why can't they take their litter home?* And, *Can't they see the big signs saying not to drop litter? Why should I pick up other people's rubbish?* And so on, until I read a caption on social media from someone who is a prolific litter picker. 'Every plastic litter you pick up is one less in the stomach of a bird, insect or animal.'

The next day I have my usual walk, breeze past a few items of litter and hesitate. As I carry on walking, I reflect on the caption and build my resolve to pick up that litter. Now, I always keep a spare folded bag with me, so on my way back I pick up the litter and pop it into the bag. I do not deliberately go looking for litter but pick up the items that cross my path along the riverbank. I chuck these in the recycling bin outside of my house before setting foot in the front door.

You should try this! Why? Because the sense of reward and satisfaction I feel is amazing. Baa used to say that it is a sin if we collude with sinners. By picking up litter and keeping it out of the bodies of other species and potentially saving their lives, you are building good karma from a relatively small gesture.

CHAPTER 13

DITCH THE BAGS

*The window of opportunity to avoid an unliveable earth
is rapidly closing. We all need to act now to make
the green revolution a reality.*

– Zoe Osmond
Director, Clean Growth UK

STEP 47 – STOP USING A PLASTIC SHOPPING BAG

I take the next step of reducing plastic by cutting down on my use of plastic shopping bags and eventually changing my shopping habits. Supermarkets don't make it easy to go totally plastic-free, but every piece of plastic avoided makes a difference to the planet.

My research on Wikipedia on plastic bags makes my chest tense up until I am gasping for breath. I hope what I share with you shall make you stop using these bags and encourage others to do the same.

Do you remember when the UK government introduced the plastic bag charge in October 2015? Something had to be done as in England alone, over 7.6 billion carrier bags were given to customers, about 61,000 tonnes in total. Just think about that for a moment!

They are showing up in oceans, up mountains and at the polar ice caps, causing major environmental damage. According to Wikipedia, plastic was discovered by accident!

'In 1933, Polyethylene, the most commonly used plastic, was created by accident at a chemical plant in Northwich, England. It was created in small batches, initially used in secret by the British military during World War II. It was then discovered that the material was industrially practical and useful. In 1965 the one-piece polyethylene shopping bag quickly began to replace cloth and paper in Europe. By 1979 it was already controlling 80 per cent of the bag market in Europe and began to spread to the United States and other countries around the world. The marketing angle was that these bags were superior to paper and other reusable bags.'[1]

In 1997, a sailor named Charles Moore discovered the Great Pacific Garbage Patch. He was the captain of a racing boat and was sailing from Hawaii to California after competing in a race. Whilst crossing the North Pacific Subtropical Gyre, Moore and his crew noticed millions of pieces of plastic surrounding his ship. He founded the Algalita Marine Research Foundation[2], and in May 2020, the Moore Institute for Plastic Pollution Research[3].

I return to Wikipedia for more info:

'In 2002 Bangladesh was the first country in the world to implement a ban on thin plastic bags, after it was found they played a key role in clogging drainage systems during disastrous flooding. Bangladesh was one of the first countries that took the bold initiative at reducing plastic, yet it is one of the worst affected countries by climate change. By 2011, one million plastic bags were consumed every minute worldwide. India hosted the #BeatPlasticPollution for World Environment Day.

'*Companies and governments around the world continue to announce new pledges to tackle plastic waste. In 2019 the EU Directive on single-use plastic products took effect as the EU aimed to lead the fight against marine litter and plastic pollution.*'[4]

Just pause and take this in. Once this sinks in, I hope you shall now be convinced to always keep a reusable non-plastic shopping bag with you. If you want to go a few steps further, you can sew tote bags from old clothes, sheets and other fabrics that are strong enough to withstand bulk shopping.

STEP 48 – STOP BUYING SANDWICH AND FREEZER BAGS

I take the next step at attempting to keep plastic out of circulation by championing reuse and refill methods over recycling. I can see that the packaging from my everyday shopping can be reused over and over again and I would only recycle as a last resort.

Prior to watching the *War on Plastic*[5] programme, my autopilot was to tear the produce (fruit/veg) out of the bag, transfer it into the fridge and chuck the plastic bag in the bin.

But the day after watching this, I stopped chucking the bread wrapper or other packaging into the bin. Now I turn it inside out, shake the crumbs out, fold it neatly and save it in my drawer for use as freezer bags.

My daughter watches me as I do this, so I explain.

'Why do I need to buy freezer bags when there are so many hard-wearing bags in my food packaging? We buy lentils, spices, fruit, vegetables and rice which come in durable plastic so I have decided not to buy sandwich and freezer bags anymore.'

I show her how I reuse a bag that contained whole cumin.

'I clean it by half filling it with water, holding the top firmly and shaking it to rinse out the residue of the contents. Then turn it inside out to dry it. I prop it here in this space between the bread bin and biscuit tin until it dries out and then pop it in the draw for the sandwich and freezer bags.'

Now, when I send food over to my children or neighbours, they get a pile of rotis in a recycled bread wrapper, or herbs from my garden in a lentil or rice bag.

The other day I passed on freshly grown parsley from my garden to my next-door neighbour.

'Ah,' she quips, 'I have fresh parsley in a brown rice bag. How wonderful.'

The irony is that I have no memory of being surrounded by plastic as a child although it is hard to imagine a life without it now. I recall my early life and the life of my previous generations. It was not like now. My parents' and grandparents' generation lived a life with no plastic, so there was and can be life without plentiful amounts of plastic.

I grew up without plastic bags. Every household had a few *thelis* – handsewn cotton bags, a version of the modern tote bag – and we took a theli with us everywhere, whether it was for shopping, taking items to another household, or for our travelling needs. We lived in a world of short supply chains that had no need for plastic. For example, we had street vendors selling their freshly grown fruits and vegetable from door to door. Most of the retailing was done by street vendors who sourced their produce from wholesalers. Word would pass around in the community when the rice, lentil, spice, saree or fabric merchant was due. For larger storage bags, such as for rice and flour, Baa used a *Gunyu* or Jute bag. The Baas in my neighbourhood often bought produce in bulk and shared it with each other.

There were no plastic food containers either. Baa used airtight stainless steel, aluminium, copper and tin containers and utensils. We used to enjoy Baa's tasty snacks, stored in these tins with her special cup of chai. We had ready access to local produce for local people, hence there was no need for fancy durable packaging from one country to another across the world contributing to plastic waste and having a colossal impact on the carbon footprint. I talk a lot more about this in my first memoir *The Best of Three Worlds*[6]. Jerome, the local smallholder, was my favourite vendor that sold goods to us.

I try to buy loose and local as best as I can and am trying to cut down on plastic packaged frozen food with the exception of peas. Why peas? Baa would buy fresh peas from Jerome and call my siblings for help with shelling them. It was always tempting to pop them in our mouths and enjoy the fresh taste but if we all did that we would miss out on the *matar bhat*, rice with peas, for dinner that night. In England you can get fresh peas occasionally although most of us rely on frozen ones because peas are a staple in Indian and English cuisine. One advantage is that the bags are very handy, so I rinse them, turn them inside out and use them for storage.

STEP 49 – STOP THE USE OF TEA BAGS FOR A SUSTAINABLE ENGLISH CUP OF TEA!

I will now switch from fruit and veg and plastic packaging to the humble cup of tea.

If there is a favourite English tradition that I have wholeheartedly embraced, then that is to enjoy my cup of tea, made the English way. It is this that brightens up my mornings and afternoons (I have two a day) and makes my time with friends and family extra special. To be handed a steaming cup of tea made the English way, to me, is a gesture of receiving love and warmth from my friends and family. Now, if you serve it to me in a pot, then you would score many extra points of my love and affection for you.

I remember in 2004, while on holiday in New York, frantically searching for my beloved English cup of tea. The Americans just do not get it because they mainly drink iced tea and coffee. I never took to coffee but took to the English cup of tea like the proverbial duck to water. In New York, I got puzzled looks when asking for a hot English cup of tea and we searched from café to café in desperation. The next time I went to America, I took a box of tea bags and just asked for hot water. This was for my daily pick-me-up as well as to promote our great British tradition.

I am, however, troubled to learn that some tea bags contain plastic! How much plastic have I accumulated in my body over my lifetime? I started using tea bags when I first came to UK over fifty years ago. At that time, I was quite in awe of the tea bag. No strainers and no mess. I have now swapped them for loose leaf tea and the use of one of those round tea strainers with a clip. It was not an easy transition by any means as I only got it right after four attempts. Twice the clip of the strainer broke, so all the loose tea tumbled into my much awaited first morning cup of tea. My devoted Hubby was not happy when he heard my woes about my morning cup of tea so he went and bought a third one after attempts to mend the clip, but the chain broke on this one. In late 2019, the search for plastic-free tea bags was fruitless. However, at the point of completing this book in May 2023, I am heartened to see plastic-free tea bags from major brands. They are using paper which still depletes the planet's natural resources so I am endeavouring to keep up with plastic and paper-free tea times with my faithful round strainer for my morning and mid-afternoon pick-me-ups.

CHAPTER 14

NO PLASTIC ON THE HOME FACE

*Once you are on this path of living an eco-friendly life,
you will discover an extraordinary journey through
a world of joy you did not even know existed*

— Hansa Pankhania

STEP 50 – USE NON-PLASTIC TOOTHBRUSHES

I am still on the 'use less plastic' mission and turn my attention to the saviour of our teeth, the trusted toothbrush.

When I was growing up, we used salt and a certain six-inch piece of twig, which we called *datan*, to clean our teeth. Before the modern toothbrush existed, people cleaned their teeth by chewing on twigs and, in some parts of the world, the practice is still common.

The twig is also called a *miswak*. I learn that in Arabic, miswak literally means a 'tooth cleaning stick.' This is trimmed from a particular species of tree. In India, this is from a Neem tree. The fibres contain sodium bicarbonate and silica, both of which are abrasive enough to help remove stains, along with natural antiseptics. In addition, the datan has a resin that supposedly forms a protective layer of essential oils over the teeth that can freshen breath.

'One of the main benefits is its convenience,' says Rahat Bashar, founder of Miswak Club[1]. 'The miswak can be used at any time of the day since no water or toothpaste is required. We've also had a lot of customers say that the main reason they use the miswak is to avoid the chemicals put in toothpaste.'

To use the twig, you have to trim the end, chew until it forms bristles, and then soak it in water to create a small brush. Every few days, you trim it again. Some studies have found that the twigs are as effective, or possibly even more effective, than using a standard toothbrush.

I ask my own dentist but this is unfamiliar to her and she says she would not personally recommend it.

'I would not rely on miswak for thoroughly cleaning teeth,' she says. 'I prefer an electric toothbrush and definitely using floss.'

I guess it would take quite a lot for a Western consumer to change to twigs for teeth cleaning, since the toothbrush and toothpaste has been used for centuries. As I am discovering on this journey, to change habits that are deeply ingrained is not easy. But it can appeal to customers looking for something a little more natural than plastic bristles so I keep on with my quest to reduce plastic.

I search the teeth hygiene sections in major stores and am disappointed to only see plastic toothbrushes wrapped in more plastic packaging. An online search uncovers other non-plastic biodegradable toothbrushes on the market which claim to have non-plastic bristles yet are crafted in a way so as not to shed the bristles and have one hundred per cent recyclable packaging. Some claim to have natural teeth whitening properties, such as the activated charcoal toothbrush which cleans the sensitive gum walls, and removes plaque as well as bad breath. I order a few of these grudgingly, because of the road miles and packing involved with an online purchase. I would much rather see them on the shelves of mainstream retail outlets.

STEP 51 – USE PLASTIC-FREE COSMETICS

We clean our teeth daily and most of us also use some form of cleansers and moisturisers or other forms of cosmetics.

One day my daughter and I decide to go shopping for plastic-free cosmetics. It turns out to be a frustrating day as we scour all the main chain stores in a major shopping mall in outer London for shampoo, face cleansers, moisturisers and deodorants.

I am in despair. It is 2019 and we are in a middle of a climate crisis, yet there is no trace of these items on the shelves of major brands. In several shops, we ask the assistants, 'Do you do shampoo bars or face cleanser bars?'

They look puzzled. 'Have you looked in the skin care aisle?' This is a common response in most of the major retailers where they signpost us from one aisle to the next. I look in the general skin care aisles where I see the usual plastic bottles and products with unnecessary plastic packaging.

After checking out products endlessly, I tell my daughter, 'I have a suspicion that most soaps and shampoos have more or less the same ingredients apart from one or two different ones according to the brand. Looking through the immense range of shampoo and shower gels, both products appear to have the same principal ingredients: surfactants and water. I can find no universal distinction, and some products sensibly claim to act as both. I gather that body wash is an emulsion/gel of water and detergent base with added functional ingredients such as moisturiser/conditioner, colourants, or fragrance and it contains milder surfactant bases than shampoos.'

We research a little more from one shelf to another and my daughter shares her thoughts.

'Hair care appears much more complicated than body care and differs according to one's hair type. It seems shampoos were created because hair becomes rough and damaged when cleansed with soap. The problem with using shampoo on the body is that some shampoo ingredients are not needed for skin and it can impart a slimy feel with the use of some.'

Our understanding therefore is that body wash and shampoo are a lot more similar than moisturisers and conditioners are. Body wash and shampoo fundamentally do the same thing – remove oil – whereas moisturiser and conditioner do not. Moisturiser literally increases the amount of water in your skin and/or creates a barrier to prevent water loss. Conditioner just makes the hair smooth; it doesn't actually add moisture.

My head is spinning as I attempt to absorb all this information and make sense of it.

We have a tea break and resume our search. Luckily, we accidentally find face cleansing bars in an independent African hair care shop and deodorant bars in another specialist independent shop on the high street.

We also visit a zero-waste refill retailer in the mall where there is a limited choice of shampoo bars and soap, and refills for cleaning products. If you were driving into a busy shopping centre, parking charges, travel miles, carrying hordes of empties for refill and lugging these all back through a busy car park would be cumbersome.

I therefore resign myself to buying small amounts by using public transport when I go there for meetings or to see friends. This may, however, not be an option for busy families with young children or elderly people with mobility issues. We need these plastic-free products on the shelves of all chain retailers on all high streets for easy access. Some refill shops also offer low-carbon delivery options if you can't make it there in person, so I resolve to look into this as a second option for future purchases.

My body wash bar and shampoo bar dilemma is resolved inadvertently by Hubby who incidentally insists on using the same brand of plastic bottled shampoo that he has been using for decades.

'Why do you have all these different soap bars everywhere in the bathroom?' he asks me with a tinge of intrigue. I have my body wash bar on the holder by the shower and shampoo bar the opposite end of the bath. He sees these as one and the same and moves them around without my knowledge. I wash my hair the next day and marvel at how soft and frizz-free it looks until I realise that I have washed it with the body wash bar! Since then, I have needed just one bar for both body and hair!

Growing up in Thika, Kenya, all family members used the same bar of soap to bath and wash hair. Nobody complained about dry hair or skin. Baa would rub coconut oil in our hair a few hours before washing and the same on dry skin. I have no memory of skin creams. These days we have different sets of plastic bottles for each family member, including children. Imagine the impact we all can make as a global community if we aim to reduce just one of these?

STEP 52 – USE PLASTIC-FREE CLEANING PRODUCTS

I encounter the same frustrations with cleaning products as was the case with cosmetics. We have a different one for different surfaces and purposes, which further adds to the amount of plastic we use and produce. When I was growing up, we used soda soap for washing clothes and other cleaning tasks. It came in long bars tied in old newspaper that Baa cut up into hand-sized usable pieces for washing clothes and dishes. For washing up dishes we would keep a bar in a stainless-steel bowl and rub an old rag on the soap and then apply it to the dirty dishes. It would easily remove stubborn grease.

When I search for sustainable cleaning products now, I find out from the Greenpeace website that it is quite simple to make your own at home. You can make effective cleaning products with household items. This is not only better for the environment but saves money and means you can upcycle old plastic bottles or jars rather than buying new. Not only is it more cost-effective to make your own, but the products also work incredibly well and, of course, you can make them smell however you want by adding essential oils, which is probably the best part. I want my home to be clean. However, at the same time, I don't really want a cocktail of harsh chemicals in the air that I breathe, or on the surfaces that I touch or prepare food on. Scientists have warned that some household cleaning products could be bad for our lungs.

All you need is white vinegar, water and your favourite essential oil. Fill a 500 ml bottle with half vinegar, half water, then add ten to twelve drops of scented oil (lavender or tea tree are ideal). Mix well and that's it! Essential oils will reduce the vinegar smell. If you are not happy with the scent, try popping some lemon/orange skins in a jar of water for a week or so. Once ready, you can add this extra liquid to the mixture to reduce, if not eradicate, the vinegar smell. If you have any tough spots of grime or dried food, then try spraying it with a liberal amount of vinegar and leaving it to soak for ten minutes before wiping. If that doesn't help, try a light sprinkling of bicarbonate of soda on the affected area, and then rub it with a damp cloth. You may want to patch test this in an inconspicuous area first to ensure that it's not going to damage the surface you are cleaning.

You can also make homemade bathroom cleaning spray, again using vinegar. This time, to the 250 ml cooled boiled water and 250 ml vinegar solution, add twenty drops of lavender oil and twenty drops of tea tree oil. Tea tree oil has excellent antiseptic and anti-fungal properties, making it one of the brilliant homemade cleaning products for tackling your bathroom.

STEP 53 – USE A SUSTAINABLE SALON

After a year of Hubby chopping my thick hair in all sorts of uneven shapes, I am able to have a professional haircut, once the Covid-19 restrictions are over.

I walk into the hairdresser's salon recommended by my neighbour. The salon is part of a 'green salon collective', salons that aim to work sustainably and ethically. The hairdresser only uses natural products with no chemicals, sublimating multi-sensory hair and scalp treatments formulated with essential oils, hydrolats and officinal extracts obtained with a biodynamic and zero-mile method. All surplus hair is recycled for compost with other products. She guides me to choose the natural, chemical-free colour for my hair and washes my hair with love and warmth. She is caring and patient when adjusting the cut, levelling my thick wavy hair, from all its battering from Hubby's pair of scissors. She has deep empathy for people and the planet and shares many tips and techniques on living an eco-friendly life with me. My heart warms up with hope and inspiration.

I leave the hairdressers with a lot more than a new hairdo. I am struck by the serendipity of this connection with the hairdresser and come away with the feeling of having won the lottery! The same feeling of peace and joy when your inner oneness merges with the outer oneness of the world.

CHAPTER 15

REFILL, REUSE, RECYCLE

If it can't be reduced, reused, repaired, rebuilt, refurbished,
refinished, resold, recycled, or composted, then it should
be restricted, redesigned or removed from production.

– Pete Seeger
Folk Singer and Social Activist

STEP 54 – REFILL AND REUSE.
ONLY RECYCLE AS A LAST RESORT

I get involved with World Refill Day[1] in June 2022. Along with our planet-protecting partners, and thousands of everyday activists around the world, this day calls on businesses, brands and governments to join the refill and reuse revolution to help make single-use plastic a thing of the past. Look out for zero-waste shops or retailers where you can refill your own containers rather than buying things in single-use packaging. Every time we make a change like this – choosing reuse over single-use – it's a step away from disposable culture and a step closer to the big refill revolution we need. Shopping in this way doesn't need to be more expensive either.

Don't worry as it can feel overwhelming at first. Choose one thing for refill you think could work for you, or the item you use the most at home such as washing-up liquid or bathroom cleaner.

Save glass jars, takeaway tubs and other containers to refill with loose products, store leftovers, or to carry a packed lunch. Download the free Refill app and find nearby places to refill your water bottle, coffee cup or lunchbox, and find plastic-free shopping options.

STEP 55 – SIGN UP TO FREECYCLE WEBSITE

The journey to switch to sustainable living is having its troughs and peaks. At times I feel good when there is progress and at other times it is frustrating, but I am keeping my commitment to carry on. I take the next step to participate in the circular economy by subscribing to Freecycle.org[2] which helps you connect with local groups and matches with people who have things they need to get rid of with people who can use them.

By using what we already have on this earth, we reduce consumerism, manufacture fewer goods and lessen the impact on the earth. Another benefit of using Freecycle.org is that it encourages us to get rid of junk that we no longer need and promotes community involvement in the process, keeping good stuff out of the landfills!

I do a mini declutter of my office and offer unwanted items on Freecycle, then share on social media using #livingwithlessplastic.

I continue to look into plastic-free products and reusable alternatives. There are multiple changes which I am trying to introduce gradually in my routines. By this time in my quest to reduce plastic, I am using eco-friendly floss, toothpaste tablets and bamboo toothbrushes. I wash my face with gentle soap bars and stop using face wipes. These are not only wasteful but they're often flushed down the toilet and end up clogging waterways. I am making my own cleaning products. I also try more environmentally friendly loo roll brands that use recycled paper.

STEP 56 – TAKE A LOCAL PLASTIC-FREE HOLIDAY

This year we swap our holiday abroad for a staycation to reduce our carbon footprint. The UK has amazing beaches, landscapes and cities to explore but the weather is not ideal if you are looking for sun, sea and sand. To overcome this we have a pre-booked break in spring and a spontaneous one in the summer when we know from the forecasts that it's going to be sunny and warm. For the second summer holiday, it has worked out well for us as it is not hard to secure a last-minute cancellation. This time we go to a beach resort in the south of England. Hubby does not want to take the train so he drives.

We are on our way when I open the snack box. It is enjoyable eating food when you are in motion in a car, and like many of you, I tuck in on short as well as long journeys. Ours has not been a household where the kitchen cupboards are crammed with bags of crisps and sugary snacks. (Of course, I keep a small supply for the odd treat.) My work has involved travel and on family travel trips, I keep containers of mixed nuts and seeds, chopped up fruit, salad and other healthy snacks such as dry fruit for the journey. There should be no excuse for not cramming in your five-a-day fruit and veg even though you are on the go.

I had a pack of plastic cutlery but after learning the facts, I now keep a supply of non-plastic cutlery as well as cups with me. This is the same when going on picnics. Non-plastic cups are bulkier to carry around and not practical if made of glass. The other day I was browsing in my bargain store and spotted a metal cup which brought up more memories of childhood. This mini jug-looking item looked quite odd and forlorn amongst all the glass and plastic ware, but for me it was an item that was part of most households in Kenya. It was durable and practical to use with cold and hot drinks.

I pick this up and pop it into my shopping bag, thinking of the times in my childhood when our African neighbour, Mrs Nyame, would offer us a drink in one of these kinds of cups while the children played together. Later I add this to my 'travelling kit' made up of non-plastic cutlery, a metal cup, a handkerchief and tea towels instead of a box of kitchen paper. I keep this in my bag (I think I need a bigger bag!) and also take it for official meetings and socialising because the UK gets through about 2.5 billion paper coffee cups a year and a million trees are felled, with almost 1.5 million litres of water used to meet this demand[3]. In light of this, every paper cup you save is definitely worth it.

It is the same with straws. Carry your metal one with you as the UK alone gets through around 4.7 billion straws each year[4]. So it's no surprise that the government banned the use of single-use plastic straws. (There are some exemptions though.) Unfortunately, the ban has not cascaded across the world so when you are out of the UK and you're offered one, just say no.

When staying overnight in a hotel, I must admit, I was guilty of picking up the hotel miniature shampoo and shower gel plastic bottles because I have been caught out with no supplies when staying at cheaper venues. I have even lined these up on my bathroom windowsill at home to enjoy some of the memories associated with these venues. I have been jolted into not doing this anymore, so I keep a small supply in reusable bottles for an emergency.

When we arrive at our South of England holiday destination, we settle into our hotel and head off to the beach nearby. It is heaving with holidaymakers, enjoying the sun, sea and sand.

I soak in the sun and the laughter and joy around me – the squeals of excited children as well as older people enjoying an ice cream quietly in a shady part of the beach. I venture up to the food stalls and the day becomes less about the sun, sea and sand. What I see adds copious rays of sunshine to my day without the help of the sun.

This is because I see visible proof of the efforts to reverse the climate and plastic crisis. It is not a common sight on my journey thus far. There is a big box where parents can drop their broken plastic buckets, spades and toys instead of leaving them on the beach. There is a sign encouraging parents to use and swap plastic beach toys. The kiosk is serving plant-based ice creams, some available in an edible cup. The next kiosk is serving vegan fish and chips and vegan sausages. It is all there. I am not having to search from one shop to the next for signs that give me hope for a sustainable earth.

I gaze at the shimmering pale blue sea, its waves waving to me from a distance. I feel the soft grainy sand on my feet and feel the rays of the sun melting by body into a relaxed state.

I am at peace, totally in love and at oneness with the world.

PART 4

CONNECTION

*It is a great feeling when positive sustainable changes are seen
to be made, but it may not be easy on this path at all times,
as you will find out from my journey.*

*Nevertheless, we will still inch forward, just by taking the first
step and gradually build the inches to feet and to yards and
eventually miles that flow effortlessly, connecting all living
beings into the vast oneness of the world.*

– Hansa Pankhania

CONNECTION WITH THE PLANET BUILDS HOPE

Eco-friendly initiatives are gathering momentum around
the globe. Connecting with these gives me hope and
reassurance that we will get through this crisis. After all, this
disconnection happened many centuries ago; therefore, we
have much ground to cover in order to reverse its detrimental
implications. Animism, the belief that all things, animals,
plants, rocks, weather systems, human behaviour and even
our words are alive and animated, started to fragment in the

17th and 18th centuries as Europeans permeating the different corners of the world viewed this as superstition and evidence of primitive backward cultures. Regrettably, this is still regarded as an embarrassment within Eurocentric cultures in some parts. Earth as a self-regulating, complex system was introduced as the Gaia theory by Lovelock and Margulis in the 1970s[1] then expanded in Lovelock's 1979 publication *Gaia: A New Look at Life on Earth*[2]. While this has educated our understanding, it has not sufficiently changed our behaviour. Animism is on the verge of a modest comeback as the world becomes more connected and self-aware.

The term 'kincentric ecology' originates from Mexico where Salmon, an academic from the 'Raramuri' people, describes it as *indigenous people viewing both themselves and nature as part of the same family that shares ancestry and origins and humans viewing their surrounding as kith and kin*. Therefore, destroying any part of this connection is believed to lead to more fragmentation and despair in the long run.

How can we hold onto hope in the face of despair? For me, reconnecting with all things living in any part of the globe reignites hope and healing. That is what will repair and restore the broken disconnected relationships. To feel hope, we must have a home where we feel safe. The concept of home, therefore, has to extend from the four walls of our physical home to the entire world.

There is growing awareness that environmental sustainability is not separate from social justice and psychosocial wellbeing as we face up to the existential crisis brought on by natural disasters and food shortages caused by climate change. In search of hope, environmentalists increasingly are searching for answers from non-Western cultures and reanimating the post-industrialised world with new narratives and images of connection and hope[3].

My quest for connection and hope for a sustainable planet happened some years ago in Scotland. For me to tell you more about this, I have to take the focus off my present abode in South England to the mesmerising stunning landscape of Scotland as I recall the train journey thirteen years ago, from Edinburgh to Inverness.

A group of my colleagues and I subscribed to a 'Sustainable Leadership Course' in Scotland in an ecovillage called Findhorn[4], which is halfway between Aberdeen and Inverness.

Findhorn is over sixty years old. The pioneering eco-settlement in the Northeast of Scotland continues to play a leading role as a research and development centre for low-carbon lifestyles, a synthesis of some of the very best of current thinking on sustainable human settlements. It is also used as a learning environment by university, school groups and professional organisations and municipalities worldwide. Today, the Findhorn ecovillage continues to prototype solutions to the regenerative design challenge of our times in the fields of food production, energy systems, built environments, biodiversity, re-localising economy and reducing our carbon footprint.

The time we spent there was enriching and enlightening. My colleagues and I immersed in an eco-friendly life for a week, joining in with community and learning activities. Unfortunately, this was at a time when other priorities in my life took over, and the learning and passion for this topic lodged itself in a hidden corner of my psyche until it ignited in 2019. As I recall the memory of that week in the ecovillage, my heart yearns for that way of living again. Alas! That is not possible due to practical reasons and the commitment of my current responsibilities. My wish is to emulate that knowledge in my current routines as much as I can and to keep learning more.

To develop more wisdom on the topic also means connecting further afield and getting an insight about what other people are doing about it in other parts of the world. With modern technology, it is easy to access and join in with global events.

CHAPTER 16

CONNECT TO CONTINUE

*Often when you think you're at the end of something,
you're at the beginning of something else.*

– Fred Rogers
US TV Host

STEP 57 – ATTEND WEBINARS AND CONFERENCES

In November 2020, I attend the Iglobalnews.com/Diwalifest[1] talk on sustainability led by a group of Indian pioneers.

I am fascinated by the concept of *Dharma*, the eternal and inherent nature of reality, regarded in Hinduism as a cosmic law underlying right behaviour and social order. This was discussed at the webinar within a business context, in the talk by Malini Mehra of Globe International legislators[2] campaigning for sustainable living. From the conversations of the panel members, I learned that the businesses that will survive are the ones that work ethically and sustainably. They predicted that hard sale, just for-profit business models will not survive as these are working against the collective consciousness. To be aligned to the collective consciousness, a business must have social responsibility and purpose. Shareholders should insist on a code of practice tied up with ethical sustainable governance. A Chief Executive must mingle with workers on the ground and share their pain as well as their gain. Businesses were advised to use mentors that offer guidance on ethical dilemmas and not just business advice aimed at profit that may be at the cost of the health of the planet. As Mahatma Gandhiji said, '*Your commerce must be for the benefit of the poor*' and '*Enough for everyone's needs, not greed.*'

The concept of *Ahimsa* – harm to others – should not happen in business either.

I too believe that work should be worship and full of actions which feeds the soul of people as well as that of the planet. Businesses can strive for a strong sense of social and planetary responsibility and can be attached to purpose and world peace, not only profits.

After the webinar that night, I woke up with a clear memory of a dream. In the dream, I found the *Upanishads*, which are ancient Hindu scriptures, from under the deep ground. I flicked through these in my dream and a page opened to an uplifting verse that whispered in my ear. I often sing this and have warm childhood memories of Baa singing it too as she went about her household tasks.

Sarve Bhavantu Sukhinah, sarve Santu Niramayah;
sarve Bhadrani Pasyantu, ma Kaschhid dukhabag Bhavet.
(May all be happy, may all be free from infirmities;
may all see good, may none partake suffering.)

In January 2021, I take part in a webinar on Conversations about Climate by Climate Psychology Alliance[3] in North America.

At this event, we are told that the physical causes of the climate crisis and humanity's complicity have been known for well over forty years. But why did we never demand that our governments act with the necessary speed? What in our psychology led us to ignore this ongoing catastrophe, and to trust that governments would safeguard our future? What forces inside ourselves held us back from doing our part to lessen our share of the damage?

My fear is that these psychological and cultural forces that encouraged us to ignore the data and move forward with 'limitless growth' will continue to hold sway, and may intensify as conditions worsen.

Sally Weintrobe, psychoanalyst, and panel member at the webinar, has thoroughly researched the answers to the question of our general paralysis in her brilliant new book, *Psychological Roots of the Climate Crisis*[4]. Her book covers topics ranging from the conflicted self to neoliberal exceptionalism, mass media and its messaging, political framing, Noah's Arkism/survivalism, and the culture of consumerism. Weintrobe argues for the paradigm shift necessary to move from the dysfunctional culture of exceptionalism, toward a culture of care that is necessary for a sustainable world.

The event included a discussion between Sally Weintrobe and Bill McKibben, another interdisciplinary luminary who has written books on this crisis, including his prophetic *The End of Nature*[5] and *Falter: Has the Human Game Begun to Play Itself Out?*[6]

After the discussion, there followed a question-and-answer period with a select group of journalists discussing the climate crisis. Sally Weintrobe went on to say that the bubble is bursting. The Covid pandemic and Black Lives Matter movement assisted in breaking down the culture of 'not caring' to a culture of care. The emphasis needs to switch from 'me' to 'us', the individual to the collective, so we come out of denial and face the truth. Young people are increasingly feeling anxious and traumatised by the implications of climate change. Bill McKibben articulated that we need to move away from a culture of selfishness of a few tycoons to one that contributes to the greater good.

My thoughts exactly!

STEP 58 – SEEK OUT MOVIES AND TV PROGRAMMES ON THE TOPIC

In March 2021, I watch Greta Thunberg documentaries[7] and the Sky News Climate Change series[8]. There are common themes that run through most of what I am watching, such as: eat less or no meat, especially beef and dairy, as cows emit methane which is worse than CO2; reduce food waste; drive and fly less; use renewable energy; make all homes and businesses more efficient; have fewer children; cut back or ditch plastic; grow more trees; and enhance biodiversity.

I also watch the movie, *Running Wild*[9], based on a true story, starring Brooke Shields, Martin Sheen and John Varty, made in 1995.

The film relates the tale of Christine Shaye, who was working for the struggling television station, Global Explorer. In an effort to find a good documentary subject, Christine travels to Africa to meet renowned documentary maker John Varty. She finds out he has been following a mother leopard for the last twelve years and is convinced she has found the best project for Global Explorer. One of the executives in the United States, however, is trying his utmost to keep the project from taking off and things in Africa aren't going that well either. John reports that the mother leopard has died after a lion attack, leaving behind her two young cubs.

John Varity describes the mother leopard he was following as his 'soul mate.' They travel from South Africa to Kenya and finally to Zambia to find the wild terrain to reintroduce the cubs to their natural habitat.

This is a beautiful rendition that reminds us that we are connected to all living beings. It shocks me when I am made aware that the conservation of natural habitat has been threatened for many decades, but we have chosen to ignore this.

The Trick[10], a BBC TV film about the climate change conspiracy to confuse people on climate decline, is inspired by true-life events that led to the Climategate scandal. Starring Jason Watkins, the film looks into the 2009 scandal that saw thousands of emails and documents leaked following a hack. It tells the story of Professor Philip Jones, the Director of Climate Research at the University of East Anglia, who was targeted in the hack. The documents were stolen by climate change deniers who argued that global warming was a conspiracy and that scientists had been hiding data.

For those who don't remember Climategate, it, too, took place just before key climate talks in Copenhagen. Hackers stole thousands of emails from the University of East Anglia's Climate Research Unit, focussing on a tiny number that seemed to suggest that scientists had manipulated data in order to exaggerate the apparent threat of climate change. They had not, as it later turned out, but the scandal still made headlines across the globe, and (it's been suggested) may have contributed to the failure of the Copenhagen talks that December.

STEP 59 – JOIN ORGANISATIONS THAT CHAMPION SUSTAINABLE LIVING

I have mentioned before that I subscribe to Greenpeace newsletters. At one point, I am researching methane gas and visit the Greenpeace website, discovering that it has a plethora of content for my sustainable path. An email pops up in my inbox. It discloses, *'Last week I emailed you asking if you could help expose Tesco's secret; they're fuelling deforestation. It's fair to say, Tesco have noticed. They quickly hid all the tweets people sent to them – but that didn't stop them from trending on Twitter.'*

Greenpeace quotes, *'Tesco don't want you to know they're the worst supermarket in the UK for forest destruction. Tesco sells more soya-fed, factory-farmed meat than any other UK supermarket – and they use one sixth of the UK's soya, 99% of it in their meat and dairy supply chain. From buying meat from companies owned by Amazon rainforest destroyers, to selling chicken and pork fed on soya from deforested land elsewhere in Brazil, Tesco must clean up their supply chain now.*

'*Help pile the pressure on Tesco and share their burning secret with your friends and family. Tesco is the UK's biggest supermarket – they have the power to make change. Help force them to act.*'[11]

I add my name to the petition and then wrangle with an ethical dilemma. Tesco is the most convenient mega store for that all-in-one place shopping as there are no independent retail outlets in my vicinity. However, there are other smaller supermarkets, Marks and Spencer, Waitrose and Lidl, so I decide to switch to these for my essentials.

STEP 60 – GET INVOLVED IN LOCAL CLIMATE CHANGE INITIATIVES

I take a trip to the Science Museum to catch the climate change exhibition, *Our Future Planet*,[12] after a meeting with a client in central London one day in August 2022. For lunch, I pop into Pret A Manger, who come across as having sustainable ethics. I offer my own cup for a hot drink. The assistant is used to this and points me in the area where other cup bringers wait for their drink. This contrasts with the assistant in McDonald's who gave me a weird, confused look when I presented my own cup.

After my trip on the train to Waterloo, I jump on the underground to get to South Kensington and walk through the bustle to join the queue for the Museum. The Science Museum visit is fascinating. The new free exhibition explores the technologies being developed to remove carbon dioxide from the atmosphere. Alongside global efforts to urgently reduce greenhouse gas emissions, scientists are racing to develop different technologies to remove and store excess carbon dioxide – the most significant cause of climate change.

Our Future Planet showcases the cutting-edge technology and nature-based solutions being developed to trap carbon dioxide released by human activity, notably the burning of fossil fuels. These include preserving ancient woodlands, and capturing carbon dioxide from the air or installing systems that prevent it leaving power stations and factories.

I find out more about the challenges involved in different solutions, the debate about what measures we need to adopt to combat climate change. How much difference could new technologies make? What can we do with the carbon dioxide after it has been captured? And why can't we just plant more trees?

I am sitting on the train back home but this time, I do not notice the world go by as busy Londoners rush home after a day's work, and tourists find their way back to their hotels. My mind is racing faster than the speed of the train. *I need to do more. I need to do more. We need to do more.* The words keep nagging me. *But how? But how?* I search all the neglected corners of my brain until I find what I am looking for. *Get active in grassroots projects.*

So the next day I do.

I research local climate change projects and as it turns out my borough council has quite a few public and corporate initiatives. It says it is committed to supporting our local flora and fauna and the diverse environments. Protecting biodiversity is one of their key objectives, playing its part locally to help halt global decline and address climate change. They have developed an online map to help discover the many wonderful green spaces on my doorstep where I can exercise, relax and enjoy nature's beauty.

But in the meantime, I subscribe to a Net Zero 360 course[13] designed by my borough to help small businesses cut down on their carbon footprint. According to the information on their site, small and medium-sized businesses make up ninety-nine per cent of all businesses in the UK so have a crucial part to play. The Net Zero 360 course is to make carbon-cutting steps as simple, easy and engaging as possible by joining fellow business pioneers on a journey that will transform their business, and hopefully our beautiful planet, for the better.

I also find out that our village has started a sustainable project, so I set up a meeting with the coordinator.

Will we be successful in turning our village into a model ecovillage to inspire the world like Findhorn? Will I go full circle and reclaim my minimalistic sustainable life like that of my childhood in Thika, Kenya? For me, it is the start of an exciting journey.

There is a new dawn in my life. I am struck by the blue in the sky as I draw the curtains in my living room. Not a single cloud wants to appear in my village today. The large veteran oak tree and silver birch beyond my bedroom window reveal their green glory. The birds gaily twitter away, celebrating another fresh youthful day.

Nature's unique magic around me mesmerise and brighten my days even more as I continue to research, write and network to take other pathbreaking actions towards the oneness of our wonderful world.

And so the journey continues…

EPILOGUE

Never doubt that a small group of thoughtful,
committed citizens can change the world.
Indeed, it is the only thing that ever has.

– Margaret Mead
American Anthropologist

How many of us realise that we are in a critical dithering phase in history? The natural world is on tenterhooks. We can destroy or save the treasures of our planet. It is up to us and the choices we make. Seeing the climate emergency unfold before us is deeply unsettling. But science tells us there is time to fix this. We can still create a world where the planet doesn't just simply recover, but it thrives. A world where entire populations have a better quality of life is still within our reach. But we'll need to work harder than ever if we're going to get there.

This decade, the 2020s, is the critical moment when we must act. Because if we wait longer, we may face a future where it's too late to keep the climate within safe levels of warming. From talking to friends and family about the climate crisis, to attending protests, to pressuring political leaders; every action that we take matters. Our contribution will help us scale up in these crucial years ahead.[1]

The physical causes of the climate crisis, and humanity's complicity, have been known for over forty years. But why did we never demand that our governments act with the necessary speed? What in our psychology led us to ignore this ongoing catastrophe, and to trust that governments would safeguard our future? What forces inside ourselves hold us back now from doing our part to lessen our share of the damage? When governments and climate change NGOs make policies and guidance or make it law, instructing people to change their behaviour, this does not mean that change will occur. It is not that easy to get results by preaching and legislation.

These psychological and cultural forces that encouraged us to ignore the data and move forward with limitless economic growth continue to hold sway and may intensify as conditions worsen. We are facing a crisis of governance as many of the world's largest economies are no longer defined by national or cultural identity and have become corporations seeking to maximise short-term profit at the expense of collateral damage. We need to move away from the culture of consumerism and bring about the paradigm shift necessary to move toward a culture of care that is necessary for a sustainable world.

To some extent, the bubble burst during the Covid pandemic, helping in a small way in breaking from a culture of not caring to a culture of care, switching the focus from 'me' to 'us' and pulling us out of denial to face the reality that we are all connected. The sad fact is that many of us lost sight of this bubble once the air molecules dissipated into thin air.

To begin this journey again may feel like a steep climb. But let me ask you this.

If you saw a magnificent mountain that was very, very tall, yet climbable, and if it was guaranteed that from its peak, you could literally see all the 8.7 million species of this earth and bask in the unique glory of this dazzling planet, would you be reluctant or revere and rejoice in each step you took as you climbed it?

The journey to sustainable living is reaching the top of that mountain by one step at a time daily.

We will have to redefine how we see ourselves and our relationships to each other and to the rest of the community of life on Earth. Only by changing our cultural narrative can we transform our vision of the future to a healthier sustainable world, and heal our relationship with life and the planet as a whole. We are creatures of habit; 'throw away' habits took years to set in so 'do not throw away' habits will take time. It is easy to give up but we can do it if we muster courage and determination to persevere for our present and future generations. Small changes make a big difference collectively. Celebrate your wins and share them to speed up the momentum.

We lead busy lives and our different circumstances can affect our ability to change. If you forget your reusable bottle or cup one day, don't feel disheartened and give up. Be proud of yourself for the other days when you do remember. It may be that you switch to a shampoo bar but need your favourite plastic bottle shampoo for a special occasion. That is fine.

As zero-waste chef Anne-Marie Bonneau says, 'We don't need a handful of people doing zero waste perfectly. We need millions of people doing it imperfectly.'[2] And if there are tips or ideas in this book that don't fit in with your current needs, that's okay. Do what you can. Write to your local Member of Parliament or join a local litter picking group or simply give up dairy for one day a week. Join forces with like-minded people in your area because system change is far and above the most effective way to tackle the climate crisis.

There is a joint responsibility. We as individuals are equally responsible for change. Governments and companies must act urgently to transform the way they run things and together, we can hold them to account and demand change, but unless we all join in and do our bit, the progress may be limited.

On a personal level, witnessing young people feeling anxious and traumatised by implications of climate change through their lives is despairing for me and many of you may also be worried about this. Therefore, it is imperative to question the oligarchs who put profits before people and the planet, and to shift from selfishness towards the greater good. I would ask them this, 'Don't you as powerful people not have grandchildren? Why are you not worried about the future of the world and climate upheavals that your future generations will face? We need you to take this on board and join in with the collective onward journey for change.'

A friend of mine came across this message from White Eagle, Hopi Indigenous Tribe in Turtle Island, and shared it in March 2020. The spiritual take of the message strongly resonates for me.

'This moment humanity is going through can be seen as a portal and as a hole. The decision to fall into the hole or go through the portal is up to you. But if you take this opportunity to look at yourself, rethink life and death, take care of yourself and others, you will cross the portal. You don't help at all by being sad and without energy. Take care of your home, take care of your body. When you are taking care of one, you are taking care of everything else. There is a social demand in this crisis, but there is also a spiritual demand. The two go hand in hand. Without the social dimension, we fall into fanaticism. But without the spiritual dimension, we fall into pessimism and lack of meaning. Establish a routine to meet the sacred every day for good things to emanate. What you emanate now is the most important thing. And sing, dance, resist through art, joy, faith and love.'[3]

These words empower me and galvanise my actions forward. Together we can turn the setback into a springboard for positive action. Spot the omens for the future and keep following the path forward. Let's work towards resetting our lives to our basic needs rather than what we want. Having unwanted material things that do not serve the natural laws of the universe has potentially offset the balance so a shift in our values has to occur in order to work towards equilibrium.

Plants and people need the same five elements – sun, water, earth, wind and space – to survive. We all have our own individual destinies, and each one is unique with the intertwining of our inner worlds and the outer. It is time to stop highlighting and bickering about our differences and come together in our common humanity and common fate of a planet in crisis. We have more in common with the natural world than that which divides us. **The decisions we make now are what will define us as who we are. What are the decisions that you are making which will define you?**

∞

I am happy, I am at one with this spectacular earth. I am connected to the soul of the world and basking in the joy of this evolving journey.

This feeling. It just does not come by itself. You have to give it to yourself and your loved ones.

Let's join hands, minds and souls, and embark on this journey together.

I would love to hear about the small or extraordinary things you have done to live an eco-friendly life after reading this book. Please let me know via www.hansapankhania.com

143

MESSAGE TO MY FUTURE GENERATIONS

Start it; you don't have to be fancy.
Keep moving; you don't have to go crazy.
Visualize; you don't have to admit it.
See the end result; it doesn't have to be material.
Expect miracles; they don't have to be huge.
Pretend you've arrived; you don't have to dance on tables.
And above all else, have fun.
This is why you started it, right?
Life, what a trip.

– Mike Dooley
Entrepreneur, Speaker
and New York Times Bestselling Author

What is the legacy that you want our generation to leave for you? Money in the bank that feeds the **consumerist culture even more**, leading to the destruction of natural habitat, or the inheritance of a thriving ubiquitous natural world and shared humanity that feeds your body, mind and spirit? Personally, I want to leave you the legacy of the richness of humanitarian values and a thriving, interconnected natural world. I want to say to you that I cared and did my best to leave a healthy flourishing planet for you.

Nature may not be natural to you, as you withdraw into more and more soulless screen time and the pernicious influences of mass media. Do you have the tenacity of weaning yourself from the allure of computers and take on the prodigious task of tuning in with the natural world? How will you preserve this planet? What is your unique talent and genius that you will share to do this? Everyone is born with a gift and gumption that is needed to fill a gap in the survival of the universe including you. **What is your gift?**

If you stay silent and make space in the visceral depth of your heart, the answer will emerge for you.

Emily Esfahani Smith in her book, *The Power of Meaning*[1], says, '*Meaning is always about connection, and about contributing to something greater than yourself. It's about loving other people and placing yourself at their service. The more you can forget yourself in the process, the greater your power to connect and to contribute, the more meaningful your life will be.*'

What gives meaning to your life?

What makes my life meaningful for me? What gets me out of bed every morning? It is not the quest for monetary acquisitions like mansions, yachts and fast cars for sure. That is not how we were brought up. My love for my family and friends certainly makes my day but there has to be a deeper reason. When I hear that what I have shared in a blog, book or verbally has made a positive difference to someone's life, this is what ignites that magical oneness with the world feeling within me. Writing about my search on how to enrich myself and the planet and saving others the trials and tribulations I have faced and sharing them with the world, makes me fulfilled with joy and the love of life.

When I share the story of my life, I see how every single event and experience has led me to be where I am today. I understand clearly how my past, present and future all connect together.

When I repeat stories from my past in my head, they can play out in a way which can sometimes appear out of context and reality but writing them down objectifies and stops them from spinning around in my head and I see everything from a new perspective.

There will be emotions that arise and come to the surface as a result of sharing my personal experiences, but I see this overall as positive. I have cried many tears whilst writing but I know that acknowledging these emotions has helped me to release anger, guilt and shame that were hiding away in the corners of my being.

This is a narrative of my battles with my conscious and subconscious processes to break some of my own habits. It is okay to say this may not be easy, and we will all have our unique challenges, but I also want to say that with a collective resolve and commitment, we can also bring about seismic change.

As my journey revealed to me, changing habits takes considerable commitment and often a grappling with our ego states. Habits can be deeply ingrained and breaking them needs dedication over a long period of time.

Sixty steps might feel a little overwhelming. It was for me when I started to write this book. However, not all the suggestions and changes I share would be the right ones for you. Instead, start with a few small things such as recycling plastic wrappers and build on these gradually to the more challenging ones, such as giving up one of the family cars. **Doing nothing is not a solution for our current generations occupying the planet**. Collectively, we can make a big difference for the natural world and hopefully you as future leaders will be thankful for the changes we make now.

When you evolve, every living being evolves around you so do not fear and hold yourselves back. I wonder about how much life we are seeping out from our beautiful planet because of our vanity and fear of judgement from others?

It is all about the power of choice: from how you choose to travel to what you choose to buy. Your lifestyle choices don't have to cost the earth. And when you influence the people in your life to make better choices too, your power is even greater.

Ana, my granddaughter, is nearly eight. She is at an age where her curiosity raises millions of questions in her mind. She is learning about different countries at school and one day announces, 'I was born in India.'

'No, Ana, you were born in England and I was born in Africa.'

'So, am I English, African or Indian?'

I explain to her that I have written *The Best of Three Worlds* ... 'When you are a few years older, you can read it and all the answers are in there.'

Similarly, when she has her own children who question her about the damage my generation has done to the planet, they can read *Best of One World*. Hopefully they can be reassured that I did my best to turn things around. Even though at times it was tough, I did keep going.

I hope you as a 'pixel' will undergo metamorphosis after reading this book, to change and join with others. As you spread the word, more people shall join in and together you will change the bigger picture.

A picture of the world and this beautiful planet where everyone is at one with oneself and at one with life on the planet. I want this oneness of the world to wrap around you making you feel snug, safe, still and at peace with all of creation.

∞

We need to move away from the culture of consumerism and bring about the paradigm shift necessary to move toward a culture of care that is necessary for a sustainable world.

— Hansa Pankhania

APPENDICES

References

PROLOGUE:

1. www.hansapankhania.com
2. Jackson, Catherine, *How we are changed*, Therapy Psychotherapist Journal, December 2020.

CHAPTER 1: WRITING FOR THE PLANET

1. Thomas, Isabel, *This Book is Not Rubbish*, Wren & Rook, 2018
2. Ibid.

PART 1: NATURE

1. positivepsychology.com/positive-effects-of-nature/

CHAPTER 2: THE LIVING MAGIC OF NATURE

1. www.naturespot.org.uk/species/mandarin
2. Coelho, Paulo, *The Alchemist*, HarperCollins, 1988
3. Ibid.

4. www.independent.co.uk/news/business/news/plants-in-the-office-boost-productivity-by-15-study-finds-9706126.html- 2014

CHAPTER 3: VEGGIE DELIGHT

1. www.countryliving.com/uk/homes-interiors/gardens/a25219874/millions-plastic-plant-pots-landfill/
2. riversidegardencentre.com/blog/broad-beans-and-the-dreaded-blackfly

CHAPTER 4: GARDEN GLORY

1. www.rhs.org.uk/gardening-for-the-environment/low-carbon-gardening/using-sustainable-materials
2. www.constructionnews.co.uk/sustainability/carbon-cementing-net-zero-22-11-2021/
3. Nex, Sally, *How to Garden the Low-Carbon Way*, RHS, 2021
4. Ibid.

CHAPTER 5: IN THE WILD

1. www.dailymail.co.uk/home/gardening/article-10961545/The-Mails-Planet-Friendly-Garden-Hampton-Court-festival-features-species-tackle-pollution.html
2. *Countryfile - Surrey Hills*, BBC TV, August 2022.
3. www.surreycc.gov.uk/culture-and-leisure/countryside/management/partnerships/heathland
4. Ibid

PART 2: WANT AND WASTE

1. www.epa.gov
2. www.quickwasters.co.uk
3. www.doximity.com/pub/shilpa-arora-md

4. www.ons.gov.uk/economy/environmental accounts/
articles/areviewofhouseholdbehaviourinrelationtofood
wasterecyclingenergyuseandairtravel/2021-11-01arev
iewofhouseholdbehaviourinrelationtofoodwasterecycl
ingenergyuseandairtravel/2021-11-01/
5. www.thedrum.com/news/2019/10/15/click-and-regret-
brits-wasting-over-half-billion-pounds-every-year-online-
unwanted
6. www.studysmarter.co.uk/explanations/environmental-
science/pollution/domestic-waste/

CHAPTER 6: FOOD, GLORIOUS FOOD

1. www.humanedecisions.com/sir-paul-mccartney-if-
slaughterhouses-had-glass-walls-everyone-would-be-
vegetarian/
2. www.nationalfoodstrategy.org/the-report/
3. www.shilpaarorand.com
4. www.fareshare.org.uk/what-we-do/hunger-food-waste/
5. www.surreyep.org.uk
6. www.sustainweb.org/news/jun21-vocal-for-local-food-
campaign/
7. Ibid.
8. Ibid.
9. www.landworkersalliance.org.uk/vocal-for-local/
10. Ibid.
11. www.sustainablefoodtrust.org/news-views/local-food-
restore-failing-food-system/
12. www.energysavingtrust.org.uk/reduce-food-waste-and-
help-environment/

CHAPTER 7: DROPS OF WATER

1. www.affinitywater.co.uk/saveourstreams

2. www.watersworthsaving.org.uk
3. www.energysavingtrust.org.uk
4. Ibid.

CHAPTER 8: MORE POWER WITH LESS ENERGY

1. www.iea.org
2. www.terna.it/en
3. www.wwf.org.uk/what-we-do/projects/reducing-carbon-emissions-uk

CHAPTER 9: CLEVER WITH CLOTHES

1. www.npr.org/2022/04/06/1091246691/microplastics-found-in-human-lungs

CHAPTER 11: THE ROAD LESS TRAVELLED

1. www.betterpoints.ltd
2. www.bikeability.org.uk
3. www.ons.gov.uk/employmentandlabourmarket/peopleinwork/employmentandemployeetypes/articles/ishybridworkingheretostay/2022-05-23
4. www.linkedin.com/in/pieter-cranenbroek/

PART 3 – PLASTIC

1. *War on Plastic*, BBC TV, June 2019

CHAPTER 12: PUT A STOP TO LITTER

1. www.thebigplasticcount.com/media/Living-with-less-plastic-handbook.pdf
2. www.greenpeace.org.uk/challenges/plastic-pollution/

CHAPTER 13: DITCH THE BAGS

1. www.wikipedia.org/wiki/Plastic_shopping_bag
2. www.algalita.org
3. www.mooreplasticresearch.org
4. www.wikipedia.org/wiki/Plastic_shopping_bag
5. *War on Plastic*, BBC TV, June 2019
6. www.hansapankhania.com

CHAPTER 14: NO PLASTIC ON THE HOME FACE

1. www.miswakclub.com

CHAPTER 15: REFILL, REUSE, RECYCLE

1. www.refill.org.uk/world-refill-day
2. www.uk.freecycle.org
3. www.theguardian.com/environment/2020/apr/26/why-britains-25-billion-paper-coffee-cups-are-an-eco-disaster
4. www.gov.uk/government/news/start-of-ban-on-plastic-straws-stirrers-and-cotton-buds

PART 4: CONNECTION

1. Lovelock, J.E., Margulis, L. (1974). *Atmospheric homeostasis by and for the biosphere: the Gaia hypothesis*. Tellus. Series A. Stockholm: International Meteorological Institute. 26 (1–2): 2–10.
2. Lovelock, James, *Gaia: A New Look at Life on Earth*, OUP, 1979
3. www.imperial.ac.uk/grantham/publications/all-publications/the-impact-of-climate-change-on-mental-health-and-emotional-wellbeing-current-evidence-and-implications-for-policy-and-practice.php
4. www.findhorn.org/about-us/

CHAPTER 16: CONNECT TO CONTINUE

1. www.iglobalnews.com
2. www.globelegislators.org/about-globe/malini-mehra
3. www.climatepsychology.us/climate-psychology-conversations
4. Weintrobe, Sally, *Psychological Roots of the Climate Crisis*, Bloomsbury Academic, 2021
5. McKibben, Bill, *The End of Nature*, Random House, 1989
6. McKibben, Bill, *Falter*, Schwartz Publishing Pty. Limited, 2019
7. www.pbs.org/show/greta-thunberg-year-change-world/
8. Skynews.com/The Daily Climate Show, March 2021 onwards
9. *Running Wild*, feature film, directed by Dee McLachlan, 1995
10. *The Trick*, BBC TV, October 2021
11. www.greenpeace.org.uk/news/five-questions-for-tesco-about-its-role-in-forest-destruction/
12. www.sciencemuseum.org.uk/what-was-on/our-future-planet
13. www.surrey.woimtg.com/?mec-events=net-zero-360-roadmap-to-your-business-net-zero-journey

EPILOGUE:

1. www.climatechangenews.com/2020/02/24/world-faces-decisive-decade-fix-global-warming-former-un-climate-chief-says/
2. www.zerowastechef.com
3. https://theglobal.school/tag/white-eagle/

MESSAGE TO MY FUTURE GENERATIONS

1. Esfahani Smith, Emily, *The Power of Meaning*, Rider, 2017

Bibliography and Further Reading

Aspey, Linda, Jackson, Catherine & Parker, Diane, (Eds), *Holding the Hope: Reviving Psychological and Spiritual Agency in the Face of Climate Change*, PCCS Books, 2023

Coelho, Paulo, *The Alchemist*, HarperCollins, 1988

Esfahani Smith, Emily, *The Power of Meaning*, Rider, 2017

Lovelock, James, *Gaia: A New Look at Life on Earth*, OUP, 1979

McKibben, Bill, *The End of Nature*, Random House, 1989

McKibben, Bill, *Falter*, Schwartz Publishing Pty. Limited, 2019

Nex, Sally, *How to Garden the Low-Carbon Way*, RHS, 2021

Thomas, Isabel, *This Book is Not Rubbish*, Wren & Rook, 2018

Weintrobe, Sally, *Psychological Roots of the Climate Crisis*, Bloomsbury Academic, 2021

TV Programmes/Films

Skynews.com/The Daily Climate Show, March 2021 onwards

Shop Well to Save the Planet, BBC TV, October 2021

Dispatches – The Truth About Nike and Adidas, Channel 4, May 2022

Big OIL vs the World, BBC TV, August 2022

Countryfile - Surrey Hills, BBC TV, August 2022

Panorama - Is the Cloud Damaging the Planet? BBC TV, February 2023

Cities - Nature's New World, BBC TV, January 2019

Our War on Plastic, BBC TV, June 2019

The Trick, BBC TV, October 2021

Running Wild, feature film, directed by Dee McLachlan, 1995

Kiss the Ground, feature film, directed by Joshua Tickell and Rebecca Harrell Tickell, 2020

Useful Websites

www.greenpeace.org

www.slowfood.org.uk

www.watersworthsaving.org.uk

www.oddbox.co.uk/blog

Activities and Resources

Many places across the UK have local groups you can join to engage in nature.

- Join the discussion on how you are connecting with nature by using the hashtag #ConnectWithNature.

- Volunteer to help on an organic farm for free accommodation and meals. WWOOF™ UK is part of the Federation of WWOOF Organisations (Worldwide Opportunities on Organic Farms) is a worldwide community that promotes awareness of ecological farming practices by providing visitors with the opportunity to live and learn about organic properties. wwoof.org.uk/

- You can access a range of activities and ideas from The Wildlife Trusts at www.wildlifetrusts.org/looking-after-yourself-and-nature.

- Learn about feeding the birds in your garden at www.cheshirewildlifetrust.org.uk/actions/how-feed-birds-your-garden

- For a more detailed guide to nature, mental health and wellbeing, you can download the Mental Health Foundation's *Thriving with Nature* guide: www.mentalhealth.org.uk/sites/default/files/Thriving-With-Nature-compressed.pdf

OTHER BOOKS BY HANSA PANKHANIA

THE BEST OF THREE WORLDS – *A soulful, cultural and historical journey across three continents.*

This is a powerful account of an exciting journey across three continents. The reader will be immersed in the fascinating fusion of three cultures and histories.

In her first memoir, Hansa talks about the values, experiences, mindfulness principles and practices that have helped her to deal with the stressful times in her life and sustain wellbeing, resilience and direction in her life.

This book will thrill and entertain by:

- Immersing you in the richness of Indian, African and English cultures.
- Educate you on the colonial history of India and Africa.
- Benefit you from East and West health techniques and philosophies.
- Enhance your life from Gandhian humanitarian values.
- Show you how to gain confidence and self-belief.

Through family, food, language, ritual and music, she uncovers the universal truths embedded in our concepts of identity. This is a unique story of how diversity can enrich our world and bring us together through the simple concept of humanity.

'I could not put this book down once I started reading it.
It draws three cultures together – African, Indian and English.
The underlying message, that we come together as one human
community wherever we are in the world, is an excellent
concept. I look forward to the author's next publication.'

J. Edwards

STRESS TO SUCCESS IN 28 DAYS – *A unique programme for total wellbeing*

This book takes you on a journey, where despite a hectic schedule, you will find the time to prevent stress in your life and propel your wellbeing so that you are bursting with energy, creativity and happiness.

Structured as a 28-day programme, the simple natural techniques taught in this book will help you:

- Learn cost-free, powerful, natural coping techniques.
- Integrate practical, easy wellbeing exercises to do in the office or at home.
- Accelerate concentration and focus.
- Enrich communication and relationships.
- Sustain healthy work-life boundaries.
- Propel resilience and a sense of relaxation.
- Boost creativity and productivity.

You do not need to make any additional time in your day as the tools in this book can be integrated within everyday activities and used throughout your lifetime to maintain a calm, productive, fulfilling and meaningful life.

'The best thing about this book is that I did not have to find any extra time. Hansa guides you to use everyday activities for reducing stress and increasing wellbeing and success. The tips are natural and simple. I found it easy to use and after a few weeks of using the techniques, now feel energised and happy.'

Elizabeth M.

STRESS TO SUCCESS STORIES – Inspiring individuals and businesses to excel

11 powerful stories covering professional and personal challenges with solutions that will ignite positive changes in your life.

This book stands out because it is not full of dry facts on stress prevention and wellbeing. This collection of workplace short stories is written from experience and inspired by true life events. The stories are drawn from Hansa's day-to-day work and contain powerful messages and coping strategies for issues such as:

Communication skills, team building, bullying, time management, mindfulness, mediation, corporate wellbeing, change management, stress management, accelerating resilience, and anger management.

The stories will inspire and empower you to:

- Overcome obstacles and challenges.
- Convince you and your organisation to prioritise wellbeing.
- Pursue your excellence and that of your business.

'Hansa skilfully integrates simple and effective techniques for managing stress with the power of storytelling.
Using a case-study style, this book shows how to deal with stressful challenges, whether in the workplace or in other areas of our lives. A very helpful book.'

S. Rowell

CHILDREN'S WELLBEING SERIES

Chakraji and Calm Callum
Chakraji and Relaxed Ravina
Chakraji and Peaceful Peter

A trilogy of wellbeing books for adults and children.

Hansa's three children's books are the tales of children's journeys to meet the magical Chakraji. The stories use imaginative storytelling to help children articulate and manage stressful situations. It's ideal for parents wanting to introduce their children to mindfulness-based stress prevention techniques. These techniques are equally helpful for adults.

The books are the first three in the series of six Chakraji books. The series introduces a range of different mindfulness-based breathing, relaxation and positive affirmation techniques, to enable children to build a repertoire of coping skills that can enhance wellbeing and resilience throughout their lives.

'Having read the previous book in the Chakraji series, my daughter was excited to try this one. She absolutely loved the pictures and the story. Chakraji is a fantastic character and Hansa does a great job of bringing her to life.
The techniques presented here are simple to use and can really help to improve the resilience of the younger generation.'

Simon Day

CHILDREN'S RELAXATION SERIES

The Chakraji Relaxation Series is a compilation of three books:

Chakraji and Beautiful Bella
Chakraji and Marvellous Mansoor
Chakraji and Special Sue Ling.

The series uses imaginative storytelling to help children articulate and manage stressful situations. This is done by introducing Chakraji, a child's magical best friend, who passes on natural breathing and mindfulness-based practical techniques that a child will find easy to use and benefit from. The aim is to enable children to build a repertoire of coping skills that can help them manage stress throughout their life.

Adults reading the story to their child will also feel calm and relaxed, meaning the book is equally helpful for the child and the reader.

The entire Chakraji series of books will help children to:

- Stay calm, grounded and focused on their education.
- Build compassion and unconditional love.
- Reduce stress, anxiety, reactivity and bad behaviour.
- Improve sleep, resilience, and well-being.
- Increase self-esteem and confidence.

'The Chakraji books are a must for children when they are having to cope with multiple challenges. The tools that are passed to the child are creative and they help a child manage moments when they are anxious. Highly recommend for parents and grandparents, and equally helpful for schools.'

David Waite

ACKNOWLEDGEMENTS

Everyone who has picked up this book and taken an interest in it.

My husband, children, grandchildren and extended family for their unconditional love and support.

My editors Jessica Powers (developmental edit), Olivia Eisinger (copy edit) and Zara Thatcher (formatting and proofreading) for their patience, wisdom and expertise.

Dr Laura Ginesi, Helen Orledge, Tim Smith, Desi Binyon, and Peter Lay for the valuable and honest feedback.

My neighbours in my resident estate for their friendship and banter.

To the community of my village for bestowing us a 'home from home.'

To the higher force beyond, for my good health and guidance with every step I take forward...

ABOUT
HANSA PANKHANIA

Hansa was born to Indian parents in Thika, Kenya, which was a British Colony in the 1950s. She came to live in the United Kingdom at the age of seventeen.

Hansa is a Coach, Speaker and published author of several Wellbeing books for adults and children. These books are available on Amazon or through her websites.

In her books, she is passionate about sharing natural wellbeing techniques, which do not cost anything but nourish your body, mind, and soul in powerful ways. Hansa and her team offer Coaching and Training on Wellbeing, Stress Prevention, Resilience Building and related topics to Individuals and Businesses.

They are committed to developing thriving ethical lifestyles including workplace cultures.

www.aumconsultancy.co.uk   www.hansapankhania.com
+44(0)7888747438

Printed in Great Britain
by Amazon

33033343R00106